THE AGRARIAN REVOLT IN WESTERN CANADA

THE AGRARIAN REVOLT IN WESTERN CANADA

A Survey Showing American Parallels

by

PAUL F. SHARP

1971
OCTAGON BOOKS
New York

Reprinted 1971
by special arrangement with the University of Minnesota Press

OCTAGON BOOKS
A Division of Farrar, Straus & Giroux, Inc.
19 Union Square West
New York, N.Y. 10003

Library of Congress Catalog Card Number: 79-154669

ISBN-0-374-97300-8

Manufactured by Braun-Brumfield, Inc.
Ann Arbor, Michigan

Printed in the United States of America

To

Rosella Anderson Sharp

Foreword

This study in Canadian-American relations is the outgrowth of a challenge issued some years ago by Professor A. L. Burt in a Canadian history seminar. He pointed out that the story of the North American West has often been approached with a national bias which ignores the parallel development of regions divided by a political boundary. These parallels, born of geography, climate, culture, and economics, have served to pour the societies of the prairie provinces and the American Northwest into much the same mold. This book analyzes the close relationship of the western Canadian farmers' movements to similar organizations in the contiguous wheat belt of the American Northwest as illustrative of the larger concept.

Canadian historians have paid scant heed to the fundamental similarities in environment that produced the same agrarian protests on both sides of the international boundary. This environment created similar problems of production, transportation, marketing, and relations with eastern metropolitan areas. Identical problems stimulated the same reactions from Canadian farmers as from their American neighbors.

American historians have often forgotten that the agricultural frontier lingered on in western Canada after it had disappeared in the United States. Thus the influence of a frontier region on American agriculture continued for many years after the American frontier had officially passed away. The forty-ninth parallel has been a far more formidable barrier to many historians than to the men and institutions they have examined.

The influence of identical environments was reinforced by the ease with which men and ideas crossed the international boundary to the north. The impact of the American republic has been great upon life in every section of the dominion, but nowhere is it more pronounced than in the Canadian West. The American farmers who helped to settle that vast region carried with them an agrarian experience which had matured under the stimuli of similar conditions in the American West. Typically western ideas quickly took root in the prairie provinces, where, combined with eastern Canadian and British traditions, they flowered in an agrarian revolt that recalls the earlier Populist crusade and parallels the contemporary agrarian movements in the American Northwest.

I am under obligation to many for assistance in this study. Visits to Canada were always a pleasure and a profit owing to the unfailing courtesy of librarians, editors, scholars, and farmers who share my enthusiasm for the history of the Canadian West. I am especially grateful to Dr. D. A. MacGibbon, chairman of the Dominion Board of Grain Commissioners, Mr. R. D. Colquette, one-time editor of the *Country Guide*, Mr. W. Norman Smith, editor of the *Western Farm Leader*, and Mr. Leonard D. Nesbitt of the Alberta Wheat Pool.

Professors Alice Felt Tyler and Rodney C. Loehr of the University of Minnesota, Professor William Lewis Morton of the University of Manitoba, and Mr. Everett E. Edwards of the United States Department of Agriculture have read the manuscript and made a number of helpful suggestions. And I am under constant obligation to Professor A. L. Burt of the University of Minnesota under whose direction this study was originally begun as a doctoral dissertation.

PAUL F. SHARP

Ames, Iowa

Contents

I

The Last Best West

The first two decades of the twentieth century were marked by one of the greatest land rushes in North American history. This was the movement into the Canadian West—called the "last best West" by those who responded to the attraction of free land. While predominantly a movement of peoples from eastern Canada and Europe, it attracted over a million Americans. The year 1890 and the "passing of the frontier" consequently are seen in a somewhat different perspective when a continental viewpoint is adopted.

The end of the nineteenth century, it is true, saw the passing of good lands within the American republic. But land hunger is a relentless force. It had ignored the international line before and was to do so for another, and perhaps last, time on this continent. Good land was to be had in Canada for the taking long after it had been exhausted south of the forty-ninth parallel. That this land lay under the Union Jack and not under the Stars and Stripes made little difference to land-hungry Americans who poured into the Canadian prairies after 1896. This movement was in the best tradition of the frontier, for the westward sweep of settlement had often shuttled across the international boundary, lured on by the existence of unoccupied lands and greater economic opportunities. Into Nova Scotia, across into the Eastern Townships of Quebec, and through Ontario the movement had swept until, diverted by the Laurentian Shield, it had turned to the south again.[1]

The great attraction of the Canadian West was land. Those who could best improve their social and economic positions by leaving their American homes were farmers, land speculators, and men

NOTE. This chapter was read at the meeting of the Agricultural History Society with the American Historical Association in New York City on December 30, 1946, and was then published in *Agricultural History*, 21:65–74 (April 1947).

[1] These movements have been carefully traced in Marcus Lee Hansen and John Bartlett Brebner, *The Mingling of the Canadian and American Peoples.*

associated with agricultural industries. To deny the agrarian nature of the American movement into the "last best West"[2] is to pose the question, Who but farmers were interested in moving in such numbers into a predominantly agricultural economy? Faltering explanations for many developments on both sides of the line have been given by those who have failed to follow the agricultural frontier as it slipped across into Canada.

The nature of a pioneer society is largely determined by the physical environment, the period of settlement, and the cultural heritage of the pioneers. Of these, the most important in many ways in the development of the prairie provinces was the geographical setting for settlement. Geographic and climatic position imposed upon the settler a narrowly seasonal occupation, and created problems of transportation and marketing which did much to shape the society he labored to build.

The familiar pattern of settlement in which the pioneer sought, as far as possible, lands similar to those left behind was followed in the last great West. Europeans, Canadians, and farmers from the American Middle West favored the Park Belt,[3] a crescent-shaped area of fertile soils with frequent "bluffs" of trees and tall grasslands stretching from the forty-ninth parallel in Manitoba to the Rockies. In this area sufficient timber, fertile black soil, and an adequate water table permitted the mixed farming familiar to farmers from the central States.

To the south and enclosed by this arc lay a bulge of the American plains—the "dry belt" or "true plains." Into this treeless, semi-arid grassland boldly pushed farmers from the Dakotas, Montana, Nebraska, and Oklahoma. For them the plains represented unlimited opportunities; they were not frightened away by the lack of timber and water.[4] The settlement of this region strikingly paralleled the early settlement of Kansas, for each occurred during a period of unusual rainfall which deceived immigrants as to the true character of the region. The rush of immigration into Alberta from 1904 to 1910 coincided with a period of rainfall adequate for wheat farming, so good, in fact, that many who had planned to irrigate their crops never connected their ditches with

[2] R. H. Coats and M. C. Maclean, *The American Born in Canada,* 11.
[3] Leo Thwaite, *Alberta, An Account of Its Wealth and Progress,* 67.
[4] W. A. Mackintosh, *Prairie Settlement, The Geographical Setting,* 1–26.

the main canals.[5] The result was the transition from ranching to farming and the confinement of the cattle industry to the Cypress Hills and the foothills of the Rockies. Later, of course, there were serious doubts as to the wisdom of breaking the plains, and numerous abandoned farms were eloquent reminders of past mistakes and forsaken hopes.[6]

Settlers in both regions fought nature on two major fronts: extreme temperatures and droughts. Frost was a spectacular enemy and the source of much unfavorable publicity for the Canadian West, but droughts proved to be the more persistent foe. The introduction of rapidly maturing wheat came too late to be an encouraging factor in settlement.[7] Dry-farming techniques developed on the American plains were of far greater significance. Though the annual precipitation seldom exceeded fifteen inches a year, the high percentage of rain during the growing season and the relatively low rate of evaporation, except in southern Alberta, fortunately gave the farmer an advantage not revealed in statistical tables.

Canadians have been puzzled by the long delay in the settlement of the Canadian West. The usual explanations for this retarded development do not tell the whole story. The Riel rebellion may have blighted the country, as some professed to believe, but it is just as reasonable to hold that settlement actually was stimulated by the numerous soldiers who, sent to crush the rebellion, saw the possibilities of the land and remained. The explanations centered upon the failure of the Canadian Pacific Railway to follow the Fleming route and its failure to send out feeder lines are only partial answers. Dominion land policy, which reserved almost half the available lands, unquestionably discouraged settlement, but it is doubtful if a more liberal policy would have made any appreciable difference in the flow of immigration. Even during the years of greatest immigration no genuine homestead philosophy was accepted.[8]

It was the existence of cheap land to the south that was largely responsible for the delay. When land prices began their upward

[5] *Manitoba Free Press,* September 5, 1919; Canada Parliament, *Sessional Papers,* 1905, no. 25, pt. 2, 78. Hereafter this source is referred to as *Sessional Papers.*

[6] Mackintosh, *op. cit.,* 124–30.

[7] Vernon C. Fowke, *Canadian Agricultural Policy, The Historical Pattern,* 236–37.

[8] *Ibid.,* 180.

spiral in the States, then the inevitable movement into the Canadian Northwest began.

The movement was inevitable in that land-hungry farmers turned to the unoccupied and promising lands to the north as soon as cheap land which did not require new and unfamiliar techniques of farming, such as irrigation, disappeared in the American West. After 1898 farmers turned in increasing numbers to this new promised land and, in terms of both population and wealth, the magnitude of the exodus to Canada impressed contemporary observers.[9] Even *Puck* magazine considered the "American invasion" significant enough to warrant comment in editorial and cartoon.

In one particularly effective cartoon, for example, *Puck* portrayed an English traveler in conversation with a "Canadian." "'Eh-ah-er,' said the Englishman, making those preparatory sounds by which the Briton warns the world of his intention to speak, 'Jolly little place this Canada — and all that sort of rot. Kindly feelings toward you Colonials too — quite so. But have you a taproom or an inn where a chap can get a pint of ale and a finnan haddie?' To which the Canadian replied, 'Well, I dunno. Up to the Washin'ton Hotel though, you kin git some Chicago ham and Milwaukee beer and New York crackers. I'm from Ioway myself — just moved North last winter — but I'm glad to see you, even if you be English. Who's king over in your country now?'"[10]

Estimates of the total number of Americans who made the trek into western Canada vary widely. Not only are immigration statistics inadequate, but the casual manner in which settlers crossed the international line without bothering with the formalities of customs must also be considered. With this in mind, Canadian officials thought that at least 20 percent of the immigrants from the States entered informally and that this percentage must be added to immigration statistics.[11] In 1914 immigration officials estimated that 925,000 persons had entered from the United States, and another estimate in the same year placed the figure at a million.[12]

[9] *Sessional Papers,* 1899, no. 13, viii.
[10] Reprinted in the *Grain Growers' Guide,* 6:7 (April 23, 1913).
[11] *Sessional Papers,* 1900, no. 13, 111; *ibid.,* 1904, no. 25, pt. 2, 95.
[12] *Ibid.,* 1914, no. 25, pt. 2, 103; J. S. Willison, "Immigrant and Settlement," in J. O. Miller, ed., *The New Era in Canada,* 109.

By 1920, therefore, the total number of immigrants probably numbered well over a million and a quarter.[13]

Even more impressive to contemporaries was the flow of wealth which accompanied the population movement across the boundary. The official estimate of wealth per capita of the settlers was one thousand dollars, but observers frequently placed the amount much higher.[14] The total wealth in cash and property taken into the Canadian West from 1900 to 1920 has been estimated to be from 270 million to a billion dollars.[15] This influx of ready capital, valuable agricultural equipment, and personal property meant that the long and arduous years necessary on most frontiers to create a comfortable living were bypassed by many settlers and were telescoped into a few years by others. The significance of this development was clearly seen by Clifford Sifton, Frank Oliver, and others interested in the development of the Canadian West. Europe could supply raw labor, but only the American West could provide men who were at once capitalists and experienced farmers. No agricultural frontier could have asked for a happier combination.

The army of farmers on the move during these years traveled chiefly by rail, though wagon trains, ox carts, and prairie schooners were in evidence. The prairie schooner, which had all but disappeared, once again became a familiar sight on the plains.[16] Since the trip could be made more cheaply by wagon than by rail, settlers heading north formed caravans comparable to those traditional in American history. A typical caravan left Crawford, Nebraska, in May 1900 with forty wagons headed for "Alberta or bust." Spare horses and cattle accompanied the caravan, and the whole scene was a "beautiful sight" to the Canadian agent who had induced the people to pull up stakes.[17] Wagons were most frequently employed by people who lived in the Dakotas or Montana, for railway connections often involved out-of-the-way routes which were expensive and time-consuming. However, settlers from Missouri, Iowa, and Kansas were known to follow the old trail along

[13] *Canadian Annual Review of Public Affairs*, 1920, 241.

[14] *Sessional Papers*, 1911, no. 25, xxix; Agnes C. Laut, "The Last Trek to the Last Frontier," *Century Magazine*, 78:99–112 (May 1909).

[15] *Canadian Annual Review of Public Affairs*, 1920, 242; Laut, *op. cit.*, 105.

[16] *Sessional Papers*, 1904, no. 25, pt. 2, 111.

[17] *Ibid.*, 1901, no. 25, pt. 2, 164.

the Missouri River to Fort Benton and then on into Canada, frequently making a journey of a thousand miles.[18]

Equally colorful were the trainloads of Canada-bound immigrants which pulled out of St. Paul, Chicago, Kansas City, and other staging areas. The completion of the "Soo Line" from St. Paul to Portal in 1894 opened the most popular route to the Northwest, and in 1900 a full train of immigrants came from as far away as Kansas City.[19] Trains leaving the Middle West often had ten to twenty carloads of stock and equipment, plus five hundred to a thousand immigrants.[20]

The influence of railroad construction on settlement and upon the pioneer economy of the Canadian prairies is well known,[21] but it is of particular interest to note that in general the settler anticipated the railroads and compelled them to follow the areas of settlement.[22] American farmers frequently struck out ahead of the railroads, many times going far beyond the distance considered economically profitable to haul grain—into the Peace River country and along the Saskatchewan River, for example.[23]

The entrance of the United States into World War I in 1917 virtually put an end to the northward movement. With the return of peace many Canadians expected to see a resumption of the immigration from the south. The desire to fill the vast lands yet open, the necessity of additional traffic for overexpanded railroads, the urgency of lessening the tax burden by the creation of additional wealth and new taxpayers—these forces impelled Canadians to look to the United States for incoming citizens.[24] Europe, they realized, needed its people for the reconstruction of a shattered continent.

However, the anticipated rush of immigrants never materialized. American farmers were too busily engaged in exploiting a brief period of prosperity to consider emigration. Moreover, avail-

[18] Hansen and Brebner, *op. cit.*, 232, quoting *Manitoba Free Press*, April 8, 1905.
[19] *Sessional Papers*, 1901, no. 25, pt. 2, 176.
[20] *Ibid.*, 1899, no. 13, pt. 2, 272; *ibid.*, 1901, no. 25, pt. 2, 175; *Manitoba Free Press*, March 31, 1919.
[21] G. P. de T. Glazebrook, *A History of Transportation in Canada*, 313–16; Mackintosh, *op. cit.*, 44–57.
[22] *Ibid.*, 55. [23] Laut, *op. cit.*, 102.
[24] Speech by the Honorable Thomas A. Crerar reported in the *Grain Growers' Guide*, 10:31 (January 24, 1917); see also *ibid.*, 10:6 (October 31, 1917).

able lands in western Canada were now largely in the hands of speculators, and land prices, though lower than in the United States, were high enough to discourage most Americans from making the change. In spite of this, Canadian real estate agents and land speculators filled newspapers with optimistic predictions of the coming land boom in which American farmers who had "more money than ever and only one way to invest it—that is, to buy more land" would pour across the boundary.[25] The sprinkling of Americans who moved into the prairie provinces was widely publicized by railway and land agents, but to no avail.[26] The mass movement was over, and the relatively small numbers who came did little but stir false hopes in Canada.[27]

The motives which impelled the last trek were of course primarily economic. The rise in the cost of land and in rents in the United States; the necessity for increased production to meet the needs of urbanization; the mechanization of agriculture which reduced the premium on labor and made possible the profitable cultivation of the vast acres of western Canada; and the decline of the fertility of older soils, resulting in increases in the costs of production, were the forces behind the movement.

The individual farmer, it is true, was not always aware of these forces, but when Dakota land sold for fifty dollars an acre in 1909 while just across the border in Saskatchewan the same type of land could be purchased for two dollars an acre, it required no knowledge of higher mathematics to visualize a substantial profit in the exchange of American for Canadian land.[28] Even more profitable was the practice of selling Iowa land for $65 to $150 an acre and then either purchasing cheap land or homesteading in Alberta or Saskatchewan. As one farmer explained it, "There is no way people can pick up $3000 easier than to come up and homestead here for three years. There will be 160 acres for nothing, and if the land is any good, at the end of the three years you can sell it for $20 an acre."[29] The result was an unhealthy specu-

[25] *Manitoba Free Press,* March 28, 1919.
[26] *Ibid.,* January 8, 1919; March 31, 1919.
[27] *Grain Growers' Guide,* 16:6 (October 17, 1923); *Manitoba Free Press,* February 1, 1919.
[28] Laut, *op. cit.,* 103.
[29] *Ibid.,* 104.

lation which frequently created a speculator more interested in a quick profit than in the permanent development of the country.

Linked to the profit motive was the very strong psychological desire to remain on the soil and keep the sons of the family on the farm.[30] With the inflation of land values in the United States, this became increasingly difficult. Moralists, politicians, and farm leaders deplored the rush to the cities and the resulting loss to the farm life of the nation, but offered no practical solutions. Many farmers found the answer by joining their neighbors in the migration north and by establishing their sons on farms adjacent to their own new homes.

There were other psychological and political motives as well. Many who made the change later gave as their reason a profound dissatisfaction with agricultural conditions in the States.[31] Professor Fred A. Shannon has advanced the suggestive hypothesis that the rise of the city acted as a "safety-valve" for rural discontent,[32] and it may well be that in a similar way, though to a lesser degree, the Canadian West served the same purpose. The more favorable agrarian legislation in the prairie provinces was a powerful magnet which attracted many disillusioned ex-Populists. Municipal hail insurance laws, a lenient land tax system, and cooperative marketing legislation were looked upon as evidences of a more progressive legislative attitude.[33]

A final and often forgotten stimulus for settlement in Canada was the appeal to racial and religious groups in the United States. It has been estimated that half of the immigrants from the United States during this period were of Canadian stock; certainly the percentage was very high.[34] To the very large number of English-speaking Canadians must also be added the French Canadians who returned to the land of their own people. Ties of blood, religion, and tradition were very strong among them, and many of them sought repatriation. Though most returned to eastern Canada,

[30] Hansen and Brebner, *op. cit.*, 223; *Manitoba Free Press*, March 31, 1919.

[31] Letter of William Nesbitt, Sr., an ex-American, *Grain Growers' Guide*, 3:15–16 (February 22, 1911).

[32] "A Post Mortem on the Labor-Safety-Valve Theory," *Agricultural History*, 19:31–37 (January 1945).

[33] *Nonpartisan Leader* (Fargo, N.D.), 5:11 (October 25, 1917).

[34] Hansen and Brebner, *op. cit.*, 227; *Sessional Papers*, 1898, no. 13, pt. 2, 111; Canada Parliament, House of Commons *Debates*, Session 1907–8, 1:1324.

there were many who joined the movement to the prairie provinces.[35]

The encouragement given to religious communities in western Canada led to a considerable migration. Many of the racial and religious communities on the Canadian prairies had their origins in the United States and not in Europe, as popularly imagined. Hutterites from South Dakota, German Lutherans from Michigan, Kentucky, and Texas, Dunkards and Mennonites from the Dakotas and Minnesota were only a few of the groups with European backgrounds who had settled in the United States but moved on when unfriendly neighbors or limited economic possibilities forced them to look for new homes.[36]

There were also striking migrations of religious communities of native-born Americans. The most notable, of course, was the migration of the Mormons from Utah into southern Alberta which began in 1887 and extended over a period of years.[37] Migrations of this type also included a Presbyterian group which left Dawes County, Nebraska, for Lacombe, Alberta, in 1900, and a large Catholic migration from St. Paul, Minnesota, to the Hoodoo plains near Prince Albert, Saskatchewan, two years later.[38]

The international character of the migration from the United States is also indicated by the number of national groups which moved on west to the prairie provinces after a short stay in the States. These included such groups as the Finnish immigrants from Minnesota who settled near Snake Lake, Alberta; Icelandic immigrants from North Dakota, Minnesota, and Washington who joined fellow countrymen in Manitoba; Norwegians and Swedes who left Minnesota and Wisconsin for northern homes; Belgians from Grosse Pointe, Michigan, who moved to Forget, Saskatchewan, and Leduc and Moranville, Alberta; and the famous German colony from Michigan under the leadership of William Richert, a former mayor of Detroit, which moved to Alameda, Saskatchewan.[39] The large numbers of persons not born in North

[35] *Ibid.*, 1909, 1:830–34; *Sessional Papers*, 1914, no. 25, pt. 2, 107.

[36] *Ibid.*, 1901, no. 25, pt. 2, 167.

[37] Donald W. Buchanan, "Mormons in Canada," *Canadian Geographical Journal*, 2:255–70 (April 1931).

[38] *Sessional Papers*, 1901, no. 25, pt. 2, 167; *ibid.*, 1904, no. 25, pt. 1, 26.

[39] Accounts of these migrations may be found in *Sessional Papers*, 1905, no. 25, pt. 2, 81; *ibid.*, 69–70; *ibid.*, 1904, no. 25, pt. 2, 97; *ibid.*, 1903, no. 25, pt. 2, 144; *ibid.*, 1914, no. 25, 106; *ibid.*, 1900, no. 13, pt. 2, 180.

America — "galvanized Americans," as Sir Sam Hughes called them[40] — who moved through the United States into Canada illustrate the dangers involved in using statistics based on "American born" to describe the nature of the movement. In such statistics about a third of the people who left American for Canadian farms are lost from view only to reappear classified under the nationality of their birth.

Though the migration into the "last best West" was the result of forces which propaganda could affect but slightly, the concerted efforts of the Canadian government and railway or land companies to educate Americans in the possibilities of the Canadian West were important. The economic desirability of settlers from the United States, as well as their farming experience, easy assimilation, and resourcefulness, led responsible authorities to make special efforts to interest them in Canadian homes.[41]

It was popularly believed on the prairies that farmers from the American plains were far more successful in meeting western conditions than immigrants from England or Europe.[42] This should occasion no surprise, for many of the English immigrants sent to the prairies were "completely innocent of rural life," and required several years to make the adjustment to their new environment.[43] Continental Europeans, unless they came from the Russian plains, were unacquainted with prairie agriculture and were difficult to assimilate into Canadian society. American immigrants, on the other hand, not only possessed the necessary technical experience but were well prepared for life in a rural society. "Most of them had been raised in the rural districts of the several states" in which Canadian agents had operated.[44] As one Canadian farmer-poet wrote, with more enthusiasm than artistry:

The Yankees in the land abound
For Uncle Sam gets all around,
And with his push and grit and go
Is sure to make the country grow.[45]

[40] House of Commons *Debates*, Session 1907–8, 4:6443.
[41] John W. Dafoe, *Clifford Sifton in Relation to His Times*, 339.
[42] *Farmers' Advocate and Home Journal*, 42:398 (March 20, 1907). Hereafter this source is cited as *Farmers' Advocate*.
[43] *Grain Growers' Guide*, 9:6 (February 16, 1916).
[44] *Sessional Papers*, 1913, no. 25, pt. 2, 91.
[45] *Grain Growers' Guide*, 4:8 (September 20, 1911).

Organized inducements for settlement took all the forms familiar to students of immigration. Canadian officials were faced with the task of convincing prospective settlers that the Northwest was not the icy land of story and poem—"Our Lady of the Snows" as Kipling had called it—but a potentially rich agricultural region capable of supporting a prosperous and happy population. This educational program was carried out through the use of advertising in more than seven thousand newspapers and farm journals, pamphlets and tracts, press excursions, and the publication of glowing reports of delegates to the West.[46]

Immigration officials quickly learned, however, that personal contacts were far more effective than any of these methods, and a highly effective system of agent coverage was developed. By 1905 salaried agents were located in most of the major cities of the Middle West, and seventy-seven subagents traveled extensively throughout the republic extolling the virtues of life in Canada.[47] Most of their activities were centered in border states, but aggressive agents early discovered that their gospel fell on responsive ears in the more remote states of Kansas, Missouri, Ohio, and Pennsylvania.[48] To encourage their efforts, head bonuses amounting to three dollars for adult males, two dollars for adult females, and one dollar for children under eighteen were paid the successful agents.[49]

To earn these bonuses, agents not only used the time-honored devices which had lured millions from Europe to North America, but also added a few profitable techniques of their own. A very effective chain-letter system was adopted to secure the names of prospective settlers, and the ultramodern magic lantern slide was pressed into use to add indisputable evidence to the agent's lecture. In smaller communities where street meetings were common, Canadian agents, with their samples of grain and photographs of prairie homesteads, competed with traveling medicine men and the Salvation Army for the Saturday night crowds. Atlases were placed in American schoolrooms; women's study clubs and cultural societies were urged to study Canada and were presented

[46] *Sessional Papers*, 1909, no. 25, pt. 2, 88; *ibid.*, 1899, no. 13, pt. 2, 106.
[47] *Ibid.*, 1906, no. 25, pt. 2, 49.
[48] *Ibid.*, 1898, no. 13, pt. 4, 82.
[49] *Ibid.*, 69.

with free "educational" literature on the subject; and church organizations and clergymen were sent special literature describing the moral and spiritual advantages of Canadian life.[50]

The most effective propaganda, however, was of a more practical nature. The brilliant record in Saskatchewan of Seager Wheeler, five times the world's champion grower of hard spring wheat, focused attention on the fertility of the Canadian plains.[51] The exhibits of Canadian crops at expositions and fairs were unusually effective, and almost every county fair in the Middle West could boast of having an attractive Canadian booth.[52] The striking number of sweepstake and championship entries in international farm expositions, dry-farming congresses, and state fairs proved to the most skeptical that the Canadian Northwest could produce the finest of crops.[53] "Letters home" describing crop prospects and actual farming conditions also induced neighbors and friends to consider a similar venture into the northland.[54]

Canadian agents frequently took advantage of population movements within the republic to hasten the movement to Canada. When lands were opened for settlement in Oklahoma Territory in 1902, for example, more farmers swarmed into the settlement than could be accommodated. Special agents were rushed to El Reno to describe the virtues of Canadian homesteads, and nearly a hundred families were convinced in a short time that Canada was the place for them.[55] Many farmers who had planned an Oklahoma sod-house homestead found themselves a few months later with a sod house in Alberta instead.

To the activities of the government were added those of railway and land companies whose profits depended upon the success of the efforts to fill the Canadian West. The Canadian Pacific Railway's plan by which a farmer could secure a "ready made" farm with a 10 percent down payment and twenty years to pay the bal-

[50] *Ibid.*, 1909, no. 25, pt. 2, 87; *ibid.*, 1899, no. 13, pt. 2, 106; *ibid.*, 1900, no. 13, pt. 2, 195; *ibid.*, 1905, no. 25, pt. 2, 31; *ibid.*, 1908, no. 25, pt. 2, 83; *ibid.*, 1905, no. 25, pt. 2, 32.

[51] *Grain Growers' Guide*, 11:5 (October 30, 1918).

[52] *Sessional Papers*, 1905, no. 25, pt. 2, 31; *ibid.*, 1914, no. 25, pt. 2, 107.

[53] *Manitoba Free Press*, July 31, 1919; *Grain Growers' Guide*, 6:15 (November 5, 1913); *ibid.*, 8:7 (October 13, 1915).

[54] *Sessional Papers*, 1899, no. 13, pt. 2, 277; Archibald S. Hurd, "Foreign Invasion of Canada," *Fortnightly Review* (London), 78:1060 (December 1902).

[55] *Sessional Papers*, 1902, no. 25, pt. 2, 146; *ibid.*, 1903, no. 25, pt. 2, 128.

ance was a very attractive offer to many hard-pressed North Dakota farmers, and the Grand Trunk Pacific and Canadian Northern railroads developed similar programs.[56] Propaganda efforts of the Saskatchewan Valley Land Company, the Canadian Land Owners Association, and the Canadian Colonization Association did much to bring to realization Canadian dreams of western empire.[57]

Organized efforts to secure American settlers for the Canadian West met with opposition both in Canada and the United States. Since the United States, unlike European nations, had no need to retain its manpower for military purposes, there was no official opposition to the Canadian immigration propaganda, but the American railway and land companies opposed the movement northward as a threat to their profits.[58]

Canadian agents in the United States felt that numerous measures were aimed at their work by American companies competing for settlers. The creation of a central publicity bureau in St. Louis by American land companies, the expenditure of three hundred thousand dollars in newspaper advertising by one American railway company to publicize its lands, the formation of the American Immigration and Minnesota Immigration associations of St. Paul, and the refusal of many bankers throughout the Middle West to lend money for investment in western Canadian land were cited as measures calculated to undermine the effectiveness of the Canadian agents' work.[59]

Counterpropaganda usually pictured the Canadian plains as a land of ice and snow, drought and disillusionment. The occasional outbursts of Canadian politicians against the loyalty of former Americans were seized upon as evidence that Americans were not wanted, and after 1914 wild stories of the hanging and shooting of aliens in Canada were circulated.[60]

Far more effective in diverting settlers from Canadian to Amer-

[56] *Nonpartisan Leader* (Fargo, N.D.), 5:11 (October 25, 1917); *Manitoba Free Press*, July 9, 1919; *Grain Growers' Guide*, 16:11 (June 20, 1923).

[57] Fowke, *op. cit.*, 180–81; Dafoe, *op. cit.*, 308; *Manitoba Free Press*, April 3, 1919; *Grain Growers' Guide*, 16:14 (August 22, 1923).

[58] *Sessional Papers*, 1905, no. 25, pt. 2, 32.

[59] *Ibid.*, 32–35.

[60] *Grain Growers' Guide*, 6:26 (March 12, 1913); *Sessional Papers*, 1917, no. 25, pt. 2, 74; *Sun* (Swift Current), January 28, 1916.

ican farms was the effect of the opening to settlement of Indian and military reservations in Minnesota, North Dakota, Montana, and Oklahoma. The Kinkaid Homestead Act of 1904 and the Enlarged Homestead Act of 1909 opened dry lands on more liberal terms than previously, and Canadian agents considered them especially damaging to their work. Other areas which attracted many who might otherwise have gone into Canada included the Florida Everglades, the Texas plains, New Mexican irrigated tracts, the Arizona plateau, the cut-over timber lands of the lake states, and the United States reclamation tracts.[61]

Opposition also came from many other directions. States especially affected by the migration took measures to halt the movement. Wisconsin barred the Canadian exhibit from its state fair in 1912 on the basis that Canadian immigration literature was "false and misleading,"[62] and in the following year the Minnesota legislature appropriated a hundred thousand dollars for immigration work. An immigration office was opened in St. Paul directly opposite the Canadian office, and attractive displays advertised lands still available within the state to the prospective settler.[63] American farm journals frequently printed adverse reports and discouraged farmers from making the change.[64] In some localities the Grange openly opposed Canadian immigration agents, and professors in agricultural colleges were charged with using their influence to obstruct the work.[65] In general, however, such opposition had little influence one way or the other.

Canadian methods of securing immigrants were also opposed by many in Canada. The use of head bonuses came under severe criticism, as did the type of literature circulated by immigration agents.[66] The *Toronto Globe* warned that "Canada cannot afford to lose her good name by rewarding her agents for practising deception," and the *Grain Growers' Guide* pointed out that "something for nothing is a mighty poor slogan upon which to develop a per-

[61] *Sessional Papers*, 1901, no. 25, pt. 2, 168; *ibid.*, 1905, no. 25, pt. 2, 34; *ibid.*, 1913, no. 25, pt. 2, 92.

[62] *Grain Growers' Guide*, 4:4 (May 29, 1912).

[63] *Sessional Papers*, 1914, no. 25, pt. 2, 112.

[64] The *Farmers' Advocate* reprinted many of these comments. See 46:1209 (August 17, 1910) for examples.

[65] *Sessional Papers*, 1915, no. 25, pt. 2, 127.

[66] House of Commons *Debates*, Session 1907–8, 1:1295; *ibid.*, 2:3313; *ibid.*, 2:3310–11; *Western Independent*, 1:3 (October 22, 1919).

manent civilization."[67] Canadian agricultural societies added their objections to literature which made conditions "appear better than they really are, believing that every immigrant should know the true facts as to conditions in this country before being induced to come here."[68] Frequent criticisms were leveled at the government policy of encouraging nationality or racial settlements on the prairies. In a resolution of the Winnipeg Board of Trade in 1911 the movement of "large numbers of Negroes from Oklahoma, USA" was condemned, because Negroes did not prove to be "satisfactory as farmers, thrifty as settlers or desirable neighbors."[69]

As a frontier society the "last best West" compared in most respects to similar regions in the American West. In the Park Belt, life assumed much the same pattern as in Manitoba, Minnesota, or Wisconsin. Log cabins usually served as the first homes and mixed farming was practiced to an increasing extent. In the "dry belt," however, the shortage of timber and water compelled the adoption of the dry-farming methods of the American plains. The sod house and miner's tent were familiar sights during the early years of settlement, and the problem of fuel was solved in traditional plains fashion.

In both regions farmers from the United States were usually successful, as revealed in countless ways. Statistically, for example, immigrants from the United States had the lowest percentage of deportees of all nationalities moving into Canada during these years. Immigration officials explained that this was true because the Americans were "mostly agriculturists in the Northwest" who possessed adequate financial resources to prevent their becoming public charges.[70]

The ease with which they adapted themselves to Canadian conditions may be explained by the very nature of the migration. American farmers possessed the necessary experience to farm successfully on the plains or in the Park Belt. This was of particular importance in the areas which required the application of dry-farming techniques. Experience gained on the American plains

[67] Quoted in House of Commons *Debates*, Session 1907–8, 2:3312–13; *Grain Growers' Guide*, 7:7 (July 8, 1914).

[68] *Ibid.*, 6:15 (April 25, 1913); *U.F.A.*, 4:5 (February 25, 1925).

[69] *Grain Growers' Guide*, 3:19 (April 26, 1911).

[70] *Sessional Papers*, 1908, no. 25, pt. 2, 135.

was extremely valuable as an example "to those from other countries who had not had the opportunity of a training in Western life."[71] Experience on the American plains had also produced such indispensable tools for western conquest as the chilled steel plow, the "dug out" for water storage, the flat furrow, and the steam plow.[72] Crops such as Grimm and Turkestan alfalfa and Turkey Red winter wheat were also brought in to add to the agricultural wealth of the Canadian West.

Along with his experience the American farmer almost always brought sufficient farm equipment and adequate capital. Thus the average immigrant from the States not only brought himself into the country without aid or subsidy, but became an immediate producer of wealth. The result was the highest percentage of farm owners of any class of immigrants coming into Canada.[73]

Though costs for getting raw land into production varied greatly throughout the Canadian West, five to fifteen dollars an acre were usually required to get a crop from new land. Regular land breakers would plow, disk, and seed the new land with steam power for an average of eight dollars an acre, and the farmer who had his own equipment could do it for considerably less.[74] A sod house or miner's tent, banked with snow during the winter season, served as a shelter for the first year or two on the plains, and in wooded regions a temporary shack could be thrown up for little or no outlay of cash. If the settler lacked the necessary equipment when he arrived, he could purchase a wagon, harrow, plow, and harness for an estimated five hundred dollars, and seed grain would require another two hundred dollars for the average 160-acre farm.[75] For those who lacked sufficient cash to get their land under crop, rapidly expanding railroads offered employment, and many settlers took advantage of this to supplement the low incomes of initial years.

The early settler in the Canadian West met the same problems

[71] *Ibid.,* 1902, no. 25, pt. 2, 143.

[72] *Ibid.,* 1910, no. 25, pt. 2, 81; Mackintosh, *op. cit.,* xi.

[73] Dafoe, *op. cit.,* 322; John Proskie, "Financial Progress of Settlers with Special Reference to the Vulcan-Lomond Area," 220, unpublished M.A. thesis, University of Alberta.

[74] John Wilson, "Experiences of a Homesteader," *Grain Growers' Guide,* 4:44–45 (December 6, 1911); Laut, *op. cit.,* 105.

[75] Wilson, "Experiences of a Homesteader," *Grain Growers' Guide,* 4:45 (December 6, 1911).

which had faced pioneer farmers across the boundary. Hail, frost, rust, grasshoppers, and drought blighted the hopes of many an optimistic settler and left him wondering why he had come north. More than one farmer's wife found that exchanging "airy nothings" with the implement agent when solid specie was in demand, and cooking for male relatives who had "bachelor's degrees in the art" were almost as trying as the vagaries of an unfriendly nature.

The inevitable ebb tide of settlement, a phenomenon so familiar in frontier history, was in full evidence by 1913. The flow of immigration into the prairie provinces dwindled and reversed itself for the same reasons that explain similar movements in other pioneer areas. Many returned to the United States with profits made by land speculation or by "mining the soil." It was popularly believed in Canada that this was the explanation for the exodus of American farmers. Stories of successful farmers who moved to California or returned to their former homes were magnified and quickly became part of the frontier tradition. That there were many who moved in "to skim the cream off of a new country" and then took vacations in California or left the prairies permanently is certainly true, but that they were more than a small minority is questionable.[76]

The return movement from Canada to the United States, which swelled in volume until the balance again favored the republic, had more basic causes than the return of those who had prospered. A succession of bad years in certain areas caused many to despair of success in the struggle against an inhospitable nature. The high cost of agricultural implements discouraged others who felt they were paying up to 20 percent more for their equipment than they had in the States and, at the same time, were receiving 10 to 20 percent less for everything they sold.[77] Even before the disastrous deflation of 1913, many were discouraged and looked elsewhere for success. The alluring stories of immigration literature and the glowing prophecies made by immigration agents were not always realized.

To many a bitter prairie settler, the cause for his failure was not mismanagement, poor farming methods, or even an unfriendly

[76] *Ibid.*, 6:3 (January 1, 1913); *ibid.*, 8:19 (December 29, 1915).

[77] J. H. Haslam, "Why Settlers Leave," *Grain Growers' Guide*, 9:8, 19, 20 (August 30, 1916).

nature, but the protectionist activities of eastern manufacturers and financiers who would make serfs of Canadian farmers.

Fundamentally, of course, the Canadian settler from the States was caught in an extremely vulnerable economy. He had moved from a national economy to an international economy which relied upon imported capital for its development and upon a world market for the sale of its only commodity. The decline of wheat prices on the Liverpool market and the failure of the London money market to pour further investments into the Canadian West brought a sharp deflation in 1913 which was keenly felt by the single crop economy. Many Americans who had come to a "pigless paradise" found that poor crops, high production costs, and lower selling prices quickly dissipated initial reserves.[78] The inevitable result was a further migration northward to regions of more rainfall, an escape to the cities, or a return to the United States.

The movement of such great numbers of American farmers into western Canada was accompanied by profound social and political consequences. Grave fears were expressed on both sides of the boundary that the migration was not an unmixed blessing. To the south there was the fear that the republic was losing many of its best farmers and much of its agricultural wealth. The *Nonpartisan Leader* deplored the loss of North Dakota's finest farmers, and Champ Clark criticized the migration as "the depleting of the Middle Western States."[79] To the north also there were serious misgivings that the flood of Americans would alter the character of Canadian society and leave it more American than British.

As early as 1903 Goldwin Smith observed that even as he wrote, "settlers from the United States are pouring into the North-West Territories, which they were sure to do when in Minnesota and Dakota land became dear. The North West will be American."[80] This unrealized prophecy became the bête noir of British Canadians who filled newspapers and periodicals with dire predictions of the future of the prairie provinces. "The great majority of these American settlers are good citizens but they have great powers of assimilation, and are first, last and always, Americans" was the

[78] *Ibid.*, 5:8 (July 17, 1912); 7:5 (January 7, 1914); 8:19 (December 29, 1915); 9:8 (August 30, 1916); 16:11 (April 25, 1923).

[79] *Nonpartisan Leader* (Fargo, N.D.), 5:11 (October 25, 1917); "Why Our Farmers Seek Canada," *Literary Digest*, 45:1217–19 (December 28, 1912).

[80] Goldwin Smith, *Reminiscences*, 417.

opinion of one westerner in 1909.[81] Another contemporary observer asked, "Must we sit by and watch one of the most promising daughter lands of the mother country being peopled by settlers of alien blood, witness the development of a policy which if not anti-British is seemingly not pro-British, and risk the political complications which may occur, in spite of all the sanguine hopes of the Dominion immigration officials?"[82]

The necessity of settling the West and the desirability of American farmers from a purely economic point of view remained unchallenged, but the admission of people unacquainted with British institutions gave rise to some alarm.[83] Lack of British interest in the development of the Canadian West was bitterly deplored by many Canadians, one of whom felt that such disinterest was "driving us into the arms of the United States, for we must get people and we must have money for our rising industries."[84]

Fears on both sides of the boundary were exaggerated and were gradually forgotten as the years passed. There were brief revivals of these apprehensions during World War I, but they vanished with the passing of the war spirit. Following the war there were expressions of anxiety that a renewed flood of American settlers and speculators would take the land needed to carry out soldier rehabilitation programs. This fear was never realized and was quickly buried under an avalanche of arguments demanding further colonization to reduce war debts and solve the financial problems of the railways and the provincial governments.[85]

In general, the ex-Americans fitted into life in the dominion without doing violence either to the society into which they moved or to the social and political ideas they brought with them. The adjustment to British institutions was made quietly and easily by the great majority, though not without certain misunderstandings and hesitations on the part of a few.

Differences in law enforcement in their new homes impressed many Americans.[86] As one farmer testified, "we have as good laws in the United States as you have in Canada, but they are not ad-

[81] John H. O'Donnell, *Manitoba As I Saw It from 1869 to Date*, 154.
[82] Hurd, *op. cit.*, 1065.
[83] House of Commons *Debates*, Session 1909, 1:853.
[84] Hurd, *op. cit.*, 1065.
[85] *Manitoba Free Press*, March 31, 1919.
[86] John Foster Fraser, *Canada As It Is*, 115.

ministered in the same way. I observe all laws on your statute books are strictly put into force, as occasion requires, while in our country many laws are a dead letter."[87] Testimony such as this repeated in letters to neighbors back home helped to create a favorable impression of this last frontier, and induced many to consider settlement in the new land.

American farmers carried with them an agrarian political experience which had a profound effect upon the political life of the prairie provinces. Reared in the atmosphere of Populism and agrarian discontent, they were well aware of the potential political power of the farmer. This experience, combined with a lack of loyalty toward traditional parties, gave the ex-Americans a political viewpoint different from anything yet experienced in Canada. "The international boundary is an invisible and intangible thing. Trade may be controlled, but ideas cannot be excluded," wrote the rancher-editor, C. W. Peterson. "Owing to the geographical situation it is almost inevitable that schemes of social reform in the two countries will go hand-in-hand."[88]

It is not surprising, therefore, to find that the ferment for agrarian reform was continental in scope. The movement of Americans into Canada stimulated the organization of Canadian branches of the Society of Equity, the Nonpartisan League, and the Ku Klux Klan, and gave strong support to the demands for prohibition, direct legislation, single tax, the recall of public officials, direct primaries, woman suffrage, free trade, and proportional representation. The Canadian West welcomed these doctrines, whatever their origin, for prairie farmers were seeking answers to the same problems that had burdened farmers in the American West twenty years earlier. This agitation and ferment set the stage for the most impressive of the third-party movements in Canada: the Progressive movement, which found its chief strength among the farmers of the prairie provinces and was another of the agrarian revolts familiar to the student of North American agriculture.

[87] O'Donnell, *op. cit.*, 122.
[88] Charles W. Peterson, *Wake Up, Canada: Reflections of Vital National Issues*, 22.

II

The Farmer's Burden

The settlement of the Canadian Northwest, and the enormous expansion in wheat production which accompanied it, marked the emergence of the premier hard spring wheat region in the world. The somewhat earlier development of the American Northwest forecast the expansion possible on the Canadian prairies. Though American immigration literature may have implied otherwise, the existence of a political boundary did not alter the conditions of nature. The climate and soils of the hard spring wheat belt were the same on both sides of the forty-ninth parallel, and the "land of the No. 1 Hard" could not be contained by an arbitrary political boundary.

Problems of transportation and marketing were much the same in both regions. The completion of the transcontinental Canadian Pacific in 1885 accomplished for the Canadian prairies what the Northern Pacific and Great Northern did for the Dakotas and Montana. Railway mileage in the prairie provinces grew from 3716 in 1901 to 11,710 in 1915.[1] Expansion in railway facilities was accompanied by an increased immigration and settlement. Encouraged by the dominion government and railway and land companies, approximately 1,500,000 settlers sought homes in the Canadian West in the fourteen years between 1900 and World War I.

Governments on both sides of the international boundary encouraged increased production through agricultural experimentation, education, and demonstration. The efforts of Canadian agricultural officials to indoctrinate prairie farmers in the importance of summer fallowing were of particular significance, though the development of Marquis wheat by Dr. Charles Saunders gained wider acclaim.

The expansion of hard wheat production in the Northwest rested

[1] Duncan A. MacGibbon, *The Canadian Grain Trade*, 40.

on the revolution in milling methods which had taken place in the 1870's and 1880's. Before 1870 hard wheat was considered less desirable than soft wheat because it was impossible to mill a clean, white flour with the flat grinding process then in use. The development of the famous "New Process" flour by Minnesota millers, however, placed the premium on the hard wheat.[2] The new techniques, which quickly spread throughout the milling industry, opened the door for the great development of the Dakotas and western Canada.

By 1900, therefore, the stage was set for the economic development of the "last best West." With the rush of settlement into the prairie provinces, another chapter in North American frontier history was written. It proved to be a story familiar to North Americans, for nature and economics imposed upon western Canadian agriculture a pattern similar to that in the American Northwest.

Economic developments in the hard spring wheat belt from 1900 until World War I were strikingly similar on both sides of the boundary.[3] The emergence of this area as a wheat-producing region coincided with the fundamental change in North American agriculture which transformed a self-sufficing economy into a commercialized economy. By the time the Dakotas, Montana, and the prairie provinces had been settled, the commercialization of agriculture was accomplished and the modern farmer had appeared. Moreover, farmers in both countries faced an increasing land cost which brought additional problems. Land values on both sides of the boundary skyrocketed in the decade after 1900 — 123 percent in Manitoba, 185 percent in Alberta, and 201 percent in Saskatchewan.[4]

Farmers who saw the dangers in this rise in property values blamed land speculation and land monopoly as its cause. Many farmers felt keen resentment against the Hudson's Bay Company, the railways, and foreign land speculators who held vast acreages from production for speculative purposes.[5] Here was a great new

[2] Charles Byron Kuhlman, *The Development of the Flour Milling Industry in the United States*, 115–19; Paul R. Fossum, "Early Milling in the Canon River Valley," *Minnesota History*, 11:271–82 (September 1930).

[3] Andrew Stewart, *The Economy of Machine Production in Agriculture*, 29.

[4] *Canadian Annual Review of Public Affairs*, 1911, 123.

[5] *Grain Growers' Guide*, 4:5 (February 21, 1912).

land opening up, and land monopolists were retarding its development. The *Grain Growers' Guide* welcomed the depression of 1913 as "a good thing for the country" if it would show the people of the West "the iniquity of speculating in land which the Creator gave for the use of His people."[6] Most farmers, however, accepted the increase in land values as a natural phenomenon and as convincing evidence of the continuing expansion of the "great Northwest."

It is now clear that much of the profit during these early years was not in wheat, but in appreciating land values. The farmers' cupidity was often aroused by advertisements which appeared in agricultural journals promoting new urban developments in the Canadian West. "Transcona: A Sure Money-Maker" fired many an acquisitive imagination, and numerous farmers who seized the "chance to get in on the ground floor" found later, when the boom collapsed, that they had burned their fingers badly. Personal experiences of this kind taught the farming community in the West that land speculation was "an evil thing for the country."

The Northwest wheat belt had an economy that rested solely on wheat. More than this, the value of the wheat was set in the Winnipeg and Minneapolis markets, which were far from the wheat fields and which were in turn responsive to the world price as established on the Liverpool market. Moreover, the functioning of this wheat economy depended upon the importation of capital, wheat being the justification for British and eastern investment in western railways and in grain marketing facilities in the Canadian Northwest as it already was in Minnesota, Montana, and the Dakotas. Thus the prairie farmer, Canadian and American, was at the mercy of forces which he keenly resented but could control only slightly.

The period from 1896 to 1920 was, in general, one of rising wheat prices and of tremendous increases in production. At the same time, however, the costs of production mounted. The farmer had to pay more for land, labor, machinery, and marketing. Surveys in the American Northwest wheat belt revealed that the producer's return was far less than popularly imagined. The rising cost of living in the United States was not due to any increase in

[6] *Ibid.*, December 24, 1913.

the farmer's net return from his products.[7] Similarly the Cost of Living Investigation in Canada in 1914–15 concluded that "it was doubtful whether during these years the wheat growers were making much more than good wages."[8] These and similar investigations confirmed the farmer's belief that he was not receiving a fair return for his labor and very little, if anything, from his investment. Few farmers denied they were getting better prices, but nearly all farmer spokesmen were convinced that agriculture was not getting its just share of the national income.

By 1909 there were grave doubts that wheat prices would continue their upward rise; in fact, fluctuations in price quotations often depressed prices to levels which were clear warnings of the disastrous declines of the future. In this year, moreover, the farmers' purchasing power reached its prewar zenith; thereafter the annual per acre value of the farmers' crops declined.[9]

The opening of the spring wheat belt coincided with the expansion of world wheat production, and during these years a balance was being reached between the demands for wheat created by urbanization and the increased production in Argentina, Australia, Canada, and the United States. Moreover, a reduction in the per capita consumption of wheat took place during this period. This helps explain the unsteadiness of the market after 1909 and its collapse in 1912–13. Before the implications of this development were fully understood, World War I intervened with its unparalleled demand for wheat production and its high prices. Before the war there were a few who saw what was happening in the world of wheat, and they sought to warn their fellow countrymen of the impending crisis in this economy. Their warnings were easily dismissed, however, as the pessimistic prophecies of men who had lost faith in the country.

Farmers on both sides of the forty-ninth parallel professed to see conditions in the other country as more favorable than in their own. American wheat producers were envious of the sympathetic governments which helped farmers on the Canadian prairies.

[7] United States Department of Agriculture *Yearbook*, 1910, 19–26; *ibid.*, 1911, 650–53.

[8] *Canada, Dominion Board of Inquiry 1914–15*, quoted by MacGibbon, *The Canadian Grain Trade*, 54.

[9] United States Department of Agriculture *Yearbook*, 1911, 651.

Direct taxes on land values, rural municipal hail insurance, and financial aid from provincial governments to farmers' grain companies were cited as evidence. American farmers also compared prices on the Winnipeg and Minneapolis exchanges and professed to see higher prices on the Canadian market.[10]

A study of the average prices on the two markets during a seventy-two-month period from 1905 to 1911, however, reveals that in only six scattered months was the Canadian price higher than the American. During 1911 and 1912 the disparity in prices so favored the American farmer that Canadians sold their grain to American elevators with a profit of nine to twelve cents per bushel despite a twenty-five-cent customs duty.[11] During 1913 the price advantage was so decisive that special measures had to be taken by United States customs officials to prevent the wholesale smuggling of cattle and grain across the border.[12] The nonexistent "lower prices paid on the Minneapolis market"[13] were used as an argument for the reform of Minneapolis marketing facilities, and as evidence of the value of farmer-owned terminals.

Canadian farmers were equally guilty of seeing imaginary advantages in the American market. They frequently advocated a sample market similar to the Minneapolis system as a "powerful factor" in remedying the evils of the grain trade in Canada. Conditions on the Minneapolis market were so good, in the Canadian view, that the system against which Dakota farmers were then in revolt was enthusiastically lauded by the *Grain Growers' Guide* as "one of the best grain marketing systems in the world."[14] Thus did distance lend enchantment.

The decade before World War I was a period of mounting discontent in the hard spring wheat belt. The agitation which grew up during these years eventually led to political revolt: to the organization of the Nonpartisan League in the Northwest states and to the Progressive movement in the prairie provinces. These political

[10] Theodore Saloutos, "The Rise of the Equity Cooperative Exchange," *Mississippi Valley Historical Review*, 32:38 (June 1945).

[11] See the experiences of J. T. Jaques, George McKitrick, and Peter Wright as recorded in *Grain Growers' Guide*, 4:20 (December 13, 1911); 4:33 (March 13, 1912); 5:5 (October 2, 1912).

[12] *St. Paul Pioneer Press*, August 21, 1913; *Grain Growers' Guide*, 6:6 (September 3, 1913).

[13] Saloutos, *op. cit.*, 38.

[14] *Grain Growers' Guide*, 4:15 (January 3, 1912).

movements, so alike yet possessed of profound differences, were born of similar underlying grievances.

In the American Northwest all phases of the wheat economy were under criticism. Railroads were attacked for charging exorbitant rates and for failure to provide sufficient cars to handle the farmers' crops. Line elevator companies were accused of underweighing, undergrading, and excessive dockage in defiance of legal proscriptions. Grain exchanges were thought to be vicious centers of gambling by grain dealers who purchased low and sold high, thus depressing the price for the farmer. Terminal elevators were charged with defrauding the farmer of his honest return by judiciously mixing high grade with low grade wheat and selling the mixture at the higher grade. Government inspectors were criticized as unsympathetic to the farmers, unduly harsh in their grading, and dominated by the grain merchants. Finally, bankers and merchants were disliked for their sharp practices which exacted heavy tolls for agricultural credits. The focal point of these attacks was the Minneapolis Chamber of Commerce, which was under constant fire for its hostile and unsympathetic treatment of the producer in the marketing system.[15]

Complaints heard on the Canadian prairies during these years were much the same, with the exception that the farmers of the north were also bitter opponents of the protective tariff.[16] American wheat farmers had been quieted by this time with the plea for "the high standard of living." These grievances, with the one exception, were accepted in both countries as explanations for the inequitable share of the national income returned to the farmer.

Grievances against the railroads first dominated agrarian agitation. Early complaints of inadequate elevator facilities, of discrimination in the location of elevator sites and disposition of cars during the shipping season, and of a failure to provide platforms for direct loading led to an investigation of the grain trade by a royal commission in 1899. The Manitoba Grain Act, which followed in 1900, secured for western Canadian farmers the protec-

[15] Theodore Saloutos, "Farmer Movements Since 1902," unpublished Ph.D. thesis, University of Wisconsin, 1940.

[16] Hopkins Moorhouse, *Deep Furrows*; Louis A. Wood, *A History of the Farmers' Movements in Canada*, 159–222; *Grain Growers' Guide*, 4:5 (February 7, 1912); B. E. Walker, "The East and West" in J. O. Miller, *The New Era in Canada*, 136.

tions which had earlier been acquired by American farmers through the Granger laws of the 1870's, the creation of the Interstate Commerce Commission in 1887, the state grain grading acts, and the railroad and warehouse commission acts of the 1890's. The enforcement of this legislation was often lax, however, and farmers on both sides of the boundary remained convinced that there existed a "silent conspiracy" to mulct them.

Complaints against transportation abuses during these years appeared chiefly in two forms—sectional rate discrimination and deliberate car shortages during the marketing season—both of which were alleged to depress wheat prices in the West. Farmer spokesmen in the American Northwest claimed with justice that southern and western producers suffered discriminatory freight rates.[17] Analyses of railroad rates at the turn of the century have revealed that Northwest wheat farmers were often paying unjustly high freight rates, and that freight costs from Dakota wheat fields to Chicago were often higher than rates from Chicago to Liverpool.[18] Remedial legislation was eventually enacted by the American Congress to remove these inequalities. The Hepburn Act of 1906 and the Mann-Elkins Act of 1910 were passed during this prewar period, and the Transportation Act followed in 1920.

Canadian wheat growers were in a more favored position. In the famous Crows' Nest Pass agreement in 1897, the dominion government granted the Canadian Pacific Railway a charter to construct a line from Lethbridge, Alberta, to Nelson, British Columbia, with a cash subsidy and a land grant. In return the railway reduced its freight charges on grain moving East to fourteen cents a hundred pounds. An agreement in 1901 between the Manitoba government and the expanding Canadian Northern system had the further effect of reducing rates between Winnipeg and the head of the lakes.[19] Thus by statutory regulation Canadian farmers paid no more than fourteen cents a hundred pounds from 1897 until the war forced a rise in freight rates. American wheat farm-

[17] H. T. Newcomb, "Changes in the Rates of Charge for Railway and Other Transportation Services," revised by Edward G. Ward, U.S. Department of Agriculture, Statistics Division, *Miscellaneous Series, Bulletin* No. 15, 21–23.

[18] John D. Hicks, *Populist Revolt,* 61; Fred A. Shannon, *The Farmers' Last Frontier: Agriculture, 1860–97,* 301.

[19] MacGibbon, *op. cit.,* 136–37; A. W. Currie, "Freight Rates in Western Canada," *Canadian Historical Review,* 21:41 (March 1940).

ers, on the other hand, were often paying twenty-five cents a hundred weight for short hauls to the Duluth or Minneapolis terminals. Low rates for grain shipments in western Canada thus placed the Canadian grain grower in a favorable position to compete with his American rival whose fields were less remote from terminal markets.

Canadian farmers found many grievances in freight rates despite the Crows' Nest Pass agreement and subsequent legislation. Though freight rates on grain could hardly be criticized with much justice, high rates on agricultural implements coming into the West caused much dissatisfaction. Prairie farmers claimed that discriminatory rates were charged on freight shipped into the prairie provinces, and cited evidence to show that comparable hauls in Ontario were less costly.[20] Statistics were marshaled to prove that rates on agricultural implements in the United States were also lower. A carload of implements shipped, for example, from Fargo to Aurelia, North Dakota, a distance of three hundred miles, cost an American farmer fifty dollars while the Canadian was forced to pay seventy-two dollars for a similar shipment from Kirkella to Colonsay, Saskatchewan, an equal distance.[21]

Rate investigations by the *Winnipeg Free Press* during 1911 confirmed many of the farmers' claims, and in 1912 a royal commission was directed to investigate freight rates. The commission made slow progress amid the confusing and contradictory evidence presented. In October 1913 the *Guide* complained that "nineteen months have elapsed since the Railway Commission began the freight rates investigation and the matter is still in the preliminary stage. How old will our grandchildren be before the West gets a square deal in freight rates?"[22] Before any adjustments could be made the war intervened and the question of rate inequalities was deferred until 1919.

The failure of railways to furnish an adequate supply of cars to handle the flow of grain during harvest was common during the early years, when western railroads still suffered from growing pains and lacked sufficient rolling stock to handle the rapidly in-

[20] *Grain Growers' Guide*, 4:5 (December 6, 1911).
[21] *Ibid.*, 4:5 (July 5, 1911).
[22] *Ibid.*, 6:7 (October 1, 1913).

creasing western yields. Farmer-critics seldom showed an understanding of the railroads' problems during these early years. This is not surprising, for failure to get the wheat to lake ports before the shipping season closed meant disastrous losses to western producers. The car shortage in 1911–12 was thought to be a "blockade," and wholly "due to the almost criminal neglect on the part of the railways to provide transportation."[23] No explanations, however plausible, could satisfy the farmers and their spokesmen, who remained convinced that the "elevator combines" and the railways had conspired to influence the government to break down the car distribution clauses of the existing grain act.

By the turn of the century farmers were less interested in measures designed to increase agricultural production and more concerned with schemes calculated to increase the return on what they marketed. In this, Northwest farmers anticipated both governments, for American and Canadian agricultural officials were as yet chiefly concerned with problems of production.

The revolt in western Canada against existing marketing conditions first appeared in the bitter criticisms of the country elevator. The complete dependence of the farmer on the line elevator as the only method of getting his wheat to market was a feature of the grain trade before the Manitoba Grain Act of 1900. This dependence frequently created a "take it or leave it" attitude on the part of elevator operators and opened the way for questionable marketing practices. Charges of undergrading, excessive dockage for dirt, weed seeds, and broken grain, and dishonest weights were too often true. The wide spread between track prices and street prices was particularly resented, for it placed the farmer who could not ship in carload lots at a costly disadvantage.[24] Many prominent leaders of the grain growers' movement, such as T. A. Crerar, E. A. Partridge, and Roderick McKenzie, felt their first resentment against the existing marketing system because of the treatment received from arbitrary elevator operators.

Western farmers were equally resentful of the role of the Winnipeg Exchange — the "House of the Closed Shutters," as E. A.

[23] *Ibid.*, 4:5 (February 7, 1912).

[24] Moorhouse, *Deep Furrows*, presents the Canadian farmers' grievances vividly and sympathetically, though not always with proper perspective.

Partridge characterized it. Remote as they were from the exchange, farmers could see few useful purposes for such an institution, and they were convinced that five grain companies controlled wheat prices to the disadvantage of the producer. Speculation and gambling depressed the price of wheat for the farmer and widened the margin of profit for the gamblers. The Northwest Grain Dealers Association, moreover, was accused of depressing the prices at the country elevators by its monopolistic control over price quotations. When the exchange expelled the Grain Growers' Grain Company from trading privileges, the farmers' criticisms grew so insistent that the Manitoba government intervened in 1908 and forced a reorganization of the exchange. Farmers rightly considered this a significant victory and a justification of much of their criticism. The marketing system was not changed, however, and the farmers' company adjusted itself with "surprising adaptability" to practices which previously had been denounced by the grain growers.[25]

Dominion grain inspectors fell under the farmers' wrath for the same reasons as state inspectors south of the boundary. Government inspection of grain was condemned as capricious, at the "whim of the inspector." The outspoken Partridge criticized the accepted methods of grain grading when he investigated the grain trade as the "farmer's representative" in 1905. Slight discolorations of the wheat berry, "a trifling number" of imperfectly formed kernels, and other subjective factors of judgment, he claimed, too often cost the farmers a heavy discount on wheat which in milling made flour of the same quality as higher graded wheat. Thousands of grain growers agreed with Partridge when he dramatically announced that "many a fraud perpetrated in a line elevator" was compounded by the "iniquities" of the inspector.[26] The grain grading system was considered too severe, and farmers were confident that Manitoba No. 2 would be graded Minnesota No. 1 just across the boundary.[27]

The fundamental importance of strict grading was misunderstood or ignored by many western farmers. They did not see that it had created something of great value to them: confidence in Ca-

[25] Harald S. Patton, *Grain Growers' Cooperation in Western Canada*, 60, 152.
[26] Wood, *op. cit.*, 60; Moorhouse, *op. cit.*, 82, 83.
[27] *Grain Growers' Guide*, 4:16 (November 15, 1911).

nadian wheat in the world market. North Dakota hard wheat enjoyed a premium on American markets because limited quantities of hard spring wheat were raised in the United States. Comparisons between the two systems were therefore limited.

Terminal elevators were also condemned for manipulating grades through grain mixing, for failing to clean the grain to grade, and for lending quantities of the farmers' wheat under their care to shippers to make up cargoes. The farmers believed there were ample opportunities at terminal facilities for graft and unethical marketing practices, and thought that "human nature being what it is, misconstructions and imputations will be made wherever there is the smallest visible motive for wrong-doing."[28] To remove temptation, the grain growers pressed for government operation of terminal facilities. Failing in this, they were eventually forced to enter terminal operation themselves, through the agency of their Grain Growers' Grain Company.

Though the major emphasis of the Canadian farmers was on marketing reform, they were beginning to question the credit and banking system during this prewar period. The farmers on the prairies were in the position of pioneers everywhere, for credit was often hard to get and usually expensive. Debts were frequently heavy, and credit needs for machinery, additional land, livestock, and working capital were great.

Farmers often naively blamed high-pressure salesmen, skilled in "the science of salesmanship," for their overbuying, but they did have legitimate grievances. High interest rates and excessive collateral were common abuses in the West. By 1910 agitation against credit discrimination was often heard on the prairies. The findings of a Saskatchewan royal commission in 1913 emphasized the need for easier credit for the grain growers and recommended provincial action.[29] The farmers' demands for financial relief were handicapped by the growing conviction that prairie agriculture was unprofitable. Few farmers realized that the louder they cried that agriculture was unprofitable, the harder it was for them to secure credit accommodation.

[28] George F. Chipman, *The Siege of Ottawa*, 25–26; *Grain Growers' Guide*, 3:5 (February 1, 1911); *ibid.*, 4:6 (February 7, 1912); speech of Peter Wright in Chipman, *op. cit.*, 26.

[29] *Grain Growers' Guide*, 6:7 (October 29, 1913).

III

"Organize! O, Organize!"

To meet the challenge of their grievances, both real and imagined, the farmers drew together. The dominant theme of the period from 1900 to 1914 was the movement of farmers into well-organized and effective societies to improve their position in the economic structure of the country. The story of the grain growers' movement, from its inception in 1901 with the formation of the Territorial Grain Growers' Association and its spread throughout the prairies, has been told so often that there is no need to repeat it here.[1] It may be worth while to point out, however, that though the grain growers' societies considered themselves "non-partisan, non-political, non-trading," they were class-conscious organizations and concerned themselves with marketing problems from the beginning.

The grain growers' movement was only one phase of a continental movement which created farmers' associations far stronger and longer lived than the colorful organizations of the previous century. The development of the grain growers' associations in Manitoba, Saskatchewan, and Alberta coincided with the organization in the United States of the Society of Equity, the Farmers' Union, the Nonpartisan League, and with the formation of local organizations which later developed into the powerful Farm Bureau movement.[2]

The rapid growth of farmers' societies during the early years of the twentieth century was facilitated by several developments. The traditional isolation of the farmer, which has received so much

[1] Louis A. Wood, *A History of the Farmers' Movements in Canada* is the most complete examination of the growth of the movement. Hopkins Moorhouse, *Deep Furrows*, Harald S. Patton, *Grain Growers' Cooperation in Western Canada*, W. A. Mackintosh, *Agricultural Cooperation in Western Canada* are the best studies.

[2] *Senate Document* No. 95, 70th Congress, 1st Session, "Cooperative Marketing"; Alfred Charles True, *A History of Agricultural Extension Work in the United States*, Department of Agriculture *Miscellaneous Publication* No. 15.

attention as a serious handicap in organization, was virtually eliminated by the advent of better roads, rural delivery of mail, the telephone, and the automobile. These innovations were of particular importance in the wheat belt, where the wheat growers' working season could be confined to seedtime and harvest. An astonishing amount of the free time remaining was used in the discussion of agricultural grievances and has been credited as a significant factor in the revolt of the wheat farmer.[3] Certainly the percentage of wheat farmers among the debaters on the courthouse steps was always very high during the off seasons.

Organization among the farmers was further aided by the high percentage of farmers who possessed experience in previous agrarian movements. Many, in both countries, had been members of the Grange or Patrons of Industry. In the United States valuable experience had been one of the heritages of the Farmers' Alliance and the Populist crusade. The Alliance had swept into western Canada, despite the youth of the country, and an even greater number of farmers joined the Patrons of Industry when it entered Manitoba in 1892.[4] Under the leadership of Charles Braithwaite, G. A. J. A. Marshall (commonly known as "Alphabetical Marshall"), John Forsyth, and H. C. Clay, the Patrons reached the respectable membership of 5000 during its brief career. A newspaper, cooperative buying and selling, and a venture into politics marked the short life of the organization on the Canadian prairies. An acrimonious controversy followed its entrance into politics, and the order quickly disintegrated.

Farmers from eastern Canada frequently had gained experience in the Grange, and settlers from the United States usually had a background in one of the American societies. English settlers, though seldom acquainted with agrarian organizations, often possessed a background in the labor or the cooperative movements which provided experience easily transferred to the prairie setting.

Every major farmers' movement organized in the United States before World War I moved into Canada in some form. The Society of Equity was of particular importance during this period, for its

[3] James E. Boyle, "The Agrarian Movement in the Northwest," *American Economic Review*, 8:506 (September 1918).

[4] H. C. Clay, "Reminiscences," *Grain Growers' Guide*, 9:21 (March 1, 1916); Wood, *op. cit.*, 124–30.

strength in Alberta was sufficient to leave an indelible impression on the farmers' organization in that province. Within five years after its organization in Indianapolis in 1902 the Equity appeared in northwestern Canada. The need for a farmers' society was keenly felt among Alberta farmers, and the Equity seemed to many to be the answer. The society grew rapidly and spread into neighboring Saskatchewan.[5] The relationship with the American society was short-lived, for the Canadians wanted their own organization and the Canadian Society of Equity was formed in 1908.

Though the Canadian society cut all formal ties with its American parent, it followed a similar program and met a similar fate. Lack of business acumen and internal dissension weakened the society in both countries. The Canadian Equity launched a series of cooperative enterprises to build elevators and to underwrite marketing organizations which proved disastrous. Ambitions outran finances, and the whole structure collapsed. As late as March 1914 shareholders were paying $7.50 a share to liquidate the Equity's unresolved liabilities.[6] This costly venture was invaluable experience to Equity men who later became leaders in the United Farmers of Alberta. The prominent U.F.A. leaders, Henry Wise Wood and George Long, were among those who lost heavily in the Equity.

The Equity, however, was not the only farmers' society in Alberta, for the stronger Alberta Farmers' Association had also been organized to end the farmers' grievances. The folly of competing for the farmers' loyalty was clearly seen, and after prolonged debate the two organizations joined forces in 1909 to form the powerful United Farmers of Alberta. The U.F.A. continued to bear testimony to its paternity, however, for it wore the official title, "The United Farmers of Alberta, Our Motto Equity."

Education was the chief emphasis of the grain growers' associations. By education they really meant indoctrination, and in this they were completely successful. The growers' societies were given financial help by the Grain Growers' Grain Company in the form of annual subsidies without which the associations could hardly have carried on their ambitious programs. In the five-year period

[5] *Grain Growers' Guide*, 10:8 (April 11, 1917); *ibid.*, 7:14 (March 18, 1914).
[6] *Ibid.*, 7:14 (March 18, 1914).

from 1909 to 1914 the growers' company poured twenty-five thousand dollars into the three provincial societies and gave another sixty thousand dollars for educational work.[7] This financial support was indispensable, and was a substantial advantage that similar agrarian societies in the States lacked. These revenues help to explain the strength of the grain growers compared with many farmer organizations south of the forty-ninth parallel, which were never able to secure sufficient money to sustain effective programs.

The organization of the dominion-wide Canadian Council of Agriculture also represented an achievement which placed Canadian agrarian societies in a more favorable position than the American organizations. Organized in 1909 to coordinate and publicize the activities of Canadian farmers, the C.C.A. performed the vital function of uniting the dominion Grange, the Manitoba and Saskatchewan grain growers, and the U.F.A. in purpose and planning. The contribution of this organization to the prestige and effectiveness of the Canadian agrarian movement had no counterpart in the United States during this period.

This does not mean, of course, that the grain growers' associations were free from strife and controversy. There were frequent revolts within the ranks of the organized farmers in every province. Annual conventions often loosed strenuous debates and bitter charges of self-aggrandizement on the part of officials.[8] Charges of inefficiency and poor judgment were frequently aired, especially against the officials of the Grain Growers' Grain Company.

In 1912 the prominent and respected grain grower, E. A. Partridge, led a revolt against T. A. Crerar's management of the company. In an open letter to the grain growers, the insurgents accused Crerar of "lack of industry and business ability," and of a failure "in carrying out the wishes of the directors."[9] The charges of the disgruntled Partridge junto were not sustained, however, and the company continued to prosper under its able manager.

In spite of frequent disagreements in policy and clashes in per-

[7] Letter of T. A. Crerar to the Manitoba Grain Growers' Association in *Grain Growers' Guide*, 7:12 (June 3, 1914).

[8] The disappearance of resolutions which never reached the floor of the conventions annoyed farmers in all societies. J. B. Musselman, the secretary of the Saskatchewan Grain Growers' Association, was frequently the target of severe criticism by farmers who resented his highhanded methods.

[9] *Grain Growers' Guide*, 4:9 (June 5, 1912).

sonality, the grain growers' movement avoided the damaging divisions which weakened the Society of Equity in the States during these years. This was due not so much to any difference in quality of leadership as to the highly controversial methods of price fixing and controlled marketing adopted by the American Society of Equity. The effort to reconcile the interests of the extremely diverse agricultural economy of the United States created complex problems for the Equity not faced by the grain growers' associations which represented the single-crop economy of wheat. The serious division of sentiment in the Society of Equity over the question of cooperation with organized labor anticipated the "broadening out" controversy that divided Canadian farmers after World War I.

The most effective organ of persuasion and indoctrination possessed by the farmers was the *Grain Growers' Guide.* This influential journal was established in 1908 and quickly became the leading farm journal in western Canada with a prewar circulation of nearly thirty thousand. The *Guide* was technically the publication of a separate corporation, but the Grain Growers' Grain Company subsidized the paper and officials of the company were on the magazine's board of directors.[10]

The influence of the *Guide* was of unusual importance because it was the official organ of the three grain growers' associations. It was an act of wisdom to make this journal the spokesman for all of the prairie societies, for this created a uniformity of opinion and a solidarity of purpose rare in farmers' movements in the United States. There the Populists, the Alliance, and earlier agrarian movements had produced scores of papers, but at best these were divided voices, lacking the impact of the single voice of the *Guide.* The closest parallels among American agrarian journals were J. A. Everitt's *Up-to-Date Farming and Gardening*, the organ of the Society of Equity, and the *Nonpartisan Leader*, the organ of the Nonpartisan League.

The similarities between the *Guide* and the *Leader* are striking. Both papers employed a directness of attack, a simplicity of argument, and an aggressive attitude that was highly effective. Each paper won the undivided loyalty of thousands of farmers, and the

[10] *Ibid.*, 6:12 (September 7, 1913).

hatred and suspicion of merchants, bankers, and industrialists. Both journals had a wide reader appeal among the rural population. Special features were included for every member of the family, along with the more serious articles that performed the "educational" function. The cartoons of John Baer in the *Leader* and Arch Dale in the *Guide* were similar in content and style, and were extremely effective in pointing out a moral to the sympathetic reader. Both papers had a strong message and both presented it effectively. The *Guide* and the *Leader* were read and believed by thousands of wheat farmers and both preached a gospel that urged the wheat belt into revolt.

Thus the farmers of western Canada were educated in the social, economic, and political aspects of grain growing. With the aid of the *Guide* the grain growers' associations performed in western Canada a great feat of indoctrination, welding the individualistic prairie farmers into an effective unit which thought and acted with uniformity. The strength born of this unity dominated the prairie governments and later held the balance of power in Ottawa. Little wonder that harassed American farmers looked across the international boundary with envy.

The parallel growth of cooperatives in the United States and Canada commands attention as an important attempt by North-American farmers to solve common marketing problems. On both sides of the boundary farmers turned increasingly to cooperative action. After a short period of inactivity following the collapse of the Farmers' Alliance, the American cooperative movement entered a period of expansion with the turn of the century.[11] The remarkable growth that followed this revival was in part due to the belief that a more equitable proportion of the nation's income would be returned to agricultural producers under a system of cooperative marketing. Cooperative organization was perhaps a normal development of the whole process of commercialization in agriculture, and it was the farmers' answer to the high degree of organization developed in industry during these years. The movement was also an expression of the crusade against monopoly which swept North America.[12]

[11] H. H. Bakken and M. A. Schaars, *Economics of Cooperative Marketing*, 47–48.

[12] See Chapter IV; G. Harold Powell, *Cooperation in Agriculture*, 10–17; H. C. Grant, "The Cooperative Movement: Developments in the Concept of Cooperation

The expansion of the movement in Canada has received the greater attention, then and since, because the farmers' grain companies were organized on a large scale and were successful much earlier in developing a fully integrated company handling the grain from the farmers' wagons to the export market. Grain cooperatives in the United States, on the other hand, were usually local in nature with 70 to 225 members.[13] It should be remembered, however, that by 1900 American farmers were well acquainted with cooperative movements and by 1907 half the farmers of America were members of some form of cooperative enterprise.[14]

In the hard spring wheat belt farmers early turned their attention to the ownership of grain-marketing facilities. Here again similar marketing problems stimulated farmers in both countries to organize. The early success of grain-selling cooperatives in the American Northwest stimulated interest in cooperation in western Canada. In 1904, for example, E. A. Partridge, the apostle of grain growers' cooperation, publicized the Minnesota Farmers' Exchange as an example of successful cooperation.[15]

The chief agency created by Canadian grain growers was the Grain Growers' Grain Company of 1906. This farmer-owned company was largely the result of the planning and dreaming of one man: E. A. Partridge of Sintaluta. The colorful career of this impetuous and idealistic farmer began in the West in 1883 on a sod-house homestead in the Qu'Appelle Valley and included such diverse experiences as frontier schoolteaching and soldiering with the Yorkton Rangers in the North-West Rebellion. When the grain growers decided to publish their own journal they selected Partridge to edit it, and when they launched their own grain company they called him to preside over its humble fortunes during the early years of struggle. But his radical views and visionary social schemes alienated many farmers, particularly as it became apparent that he was intolerant of criticism and reluctant to accept advice, and he lost his leadership to younger, less outspoken men.

The Sintaluta farmer preached his gospel of cooperation with

in Western Canada," *Canadian Journal of Economics and Political Science*, 3:412–20 (August 1937).

[13] Powell, *op. cit.*, 122.

[14] Fred A. Shannon, *The Farmer's Last Frontier: Agriculture, 1860–1897*, 346.

[15] Mackintosh, *op. cit.*, 18.

such passion and persistence that his friends and neighbors joined in organizing the capital stock venture to market their grain. Though shares in the new company sold for only twenty-five dollars, the struggling organization experienced such difficulty selling them that the Canadian novelist, Hopkins Moorhouse, was later inspired to recall: "That stock-selling campaign will dodder through to the Final Memory of those who took part in it. The man who stood on the street-corner and offered ten-dollar gold-pieces for a dollar had no harder task. Blood from stones! Milk from dry cows! Although ten per cent was all the cash that was asked apparently some farmers were so hard up that if yarn were selling at five cents per mile, they couldn't buy enough of it to make a pair of mitts for a doodlebug!"[16]

The dramatic story of the company's struggle for existence against the ill will and organized opposition of the private grain dealers has often been told,[17] but it should be pointed out here that the boycotts, expulsions from the exchange, and credit restrictions suffered by the Canadian company were similar to the struggles which had threatened the farmers' grain companies in the United States at the turn of the century.[18] Government intervention eventually protected the farmers' interests in both countries. In 1912 the growers' company in Canada leased one hundred and thirty-five elevators from the Manitoba government after an attempt by provincial authorities to operate public elevators on the farmers' behalf had failed. With the financial support of their provincial governments, grain growers in the other prairie provinces also organized elevator systems: the Saskatchewan Cooperative Elevator Company in 1911 and the Alberta Farmers' Cooperative in 1913.

The formation of these companies left unsolved the problem of terminal elevator facilities, as had the organization of local grain cooperative associations in the United States. Many farmers were convinced that abuses at terminal elevators could be eliminated only through government monopoly of these facilities.[19] Failure to secure government operation led to the entry of the Grain Growers'

[16] Moorhouse, *op. cit.*, 91.
[17] Moorhouse, *op. cit.*, and Patton, *op. cit.*, have the best accounts.
[18] *Senate Document* No. 95, 70th Congress, 1st Session, 56–60.
[19] Theodore Saloutos, "The Rise of the Equity Cooperative Exchange," *Mississippi Valley Historical Review*, 32:38 (June 1945); Patton, *op. cit.*, 147.

Grain Company into terminal operation in 1912 with the leasing of Canadian Pacific Railway terminal B at Fort William.[20]

Similar efforts to release American wheat growers from the domination of private grain companies at terminal exchanges were taken through the organization of the Equity Cooperative Exchange. Though formed in 1908, it was not until three years later that the Equity Exchange began operations. This precipitated the bitter struggle with the Minneapolis Chamber of Commerce, which feared the Equity Cooperative Exchange might grow to the dominant stature of the Grain Growers' Grain Company in Canada.[21]

The organization of cooperative grain associations met determined opposition from private grain dealers in both countries. The struggle of the early farmers' grain companies in the United States foreshadowed the dramatic struggles which were to follow in Canada when the grain growers' company entered the Winnipeg Exchange. The price wars, boycotts, credit refusals by unsympathetic banking houses, and the organized opposition of line elevator companies and grain exchange members led to the organization of statewide farmers' grain-dealer associations in the United States. These were the same methods the Canadian grain dealers employed to eliminate the Grain Growers' Grain Company from competition, and the effect of the opposition was much the same in both countries. Farmers rallied to the support of their threatened companies and through political intervention forced the exchanges to accept the principle of grain marketing by farmer-owned companies.[22]

Comparisons are difficult, if not impossible, for statistics of cooperative activity during this period are inadequate, and many "farmer-owned" companies were not cooperatives but capital stock companies organized by grain growers. To indicate the parallel growth, however, it is possible to make limited comparisons.

By 1914, for example, there were 264 cooperative elevator companies in North Dakota, 241 in Minnesota, 228 in Iowa, and 183 in Nebraska. In the prairie provinces in the same year, there were 135 farmers' elevators in Manitoba, 192 in Saskatchewan,

[20] *Ibid.*, 148.

[21] Saloutos, *op. cit.*

[22] *Senate Document* No. 95, 70th Congress, 1st Session, 53–62; Patton, *op. cit.*, 62–76.

and 52 in Alberta.[23] To the operation of these elevators, of course, must be added the wheat shipped directly to the Grain Growers' Grain Company in Winnipeg. Membership in the American grain cooperatives in 1911 numbered 225,000 and by 1915 one in four farmers in North Dakota and one in five in Minnesota were members of grain cooperatives.[24]

Another measure of the extent of cooperative grain marketing was that submitted in a study made for the United States Federal Trade Commission, which discovered that during the five crop years beginning in 1912, 23.7 percent of the grain elevators in North Dakota, 21.34 percent in Minnesota, 23.9 percent in Montana, and 23.08 percent in South Dakota were cooperatively owned.[25] To the north farmer-owned cooperatives controlled more than 20 percent of the elevators in the three prairie provinces.[26]

It is perfectly clear, therefore, that by 1914 farmer-owned grain companies were an accepted feature of the grain-marketing system. Though their influence was more strongly felt in the Canadian West, farmers' companies were sufficiently numerous in the United States to prevent a repetition of the conditions that had prevailed in the 1890's.[27]

The major concern of western Canadian farmers before World War I was the demand for a lower tariff. Before the Manitoba Grain Act and the court decisions which sustained it, farmers were chiefly concerned with securing remedial legislation in the grain trade. By 1905, however, the chief emphasis of the grain growers' agitation was focused on the crusade to reduce the tariff walls that the National Policy had erected.

The movement against the protective tariff was motivated by principle and by self-interest. The Canadian West was the home of many English free-traders who had been indoctrinated in the efficacy of free trade in the mother country. Their convictions were confirmed when they analyzed the economy into which they had

[23] O. B. Jesness and W. H. Kerr, "Cooperative Purchasing and Marketing Organizations among Farmers in the United States," Department of Agriculture *Bulletin* No. 547, 13; Patton, *op. cit.*, 96; Wood, *op. cit.*, 318, 322.

[24] Powell, *op. cit.*, 122; Robert H. Bahmer, "The Economic and Political Background of the Nonpartisan League," 156, unpublished Ph.D. thesis, University of Minnesota, 1941.

[25] *Senate Document* No. 95, 70th Congress, 1st Session, 52.

[26] Mackintosh, *op. cit.*, 73.

[27] Bahmer, *op. cit.*, 164.

moved, for the Canadian West was forced to sell in an unprotected world market and buy in a protected home market. This disadvantage was keenly felt in the West and was commonly believed to be the explanation for the slow development of the prairies. As one contemporary observer saw it, "If the West fills up, the tariff will go, and if the tariff does not go, the West will not fill up."[28]

It is interesting that eastern protectionist newspapers sought to discredit the free trade movement on the prairies by characterizing it as "led by a bunch of American agitators." This charge, commonly used in Canada to frighten the sensitive, was statistically false and gave credit where credit was not due.[29] Few ex-Americans possessed sufficient background in tariff questions to play any significant role in the matter. The tariff issue among farmers in the States had long been subordinated to other issues, such as the currency and credit questions and railway and marketing controls. In fact, most Americans had difficulty in making the adjustment to the international economy of the Canadian prairies after leaving the shelter of a national economy that consumed the great bulk of the national production. Thousands of British-born grain growers, on the other hand, had come from an economy which had practiced free trade, and they quickly transferred Old World experience to the New World setting.

Western farmers fought the protective tariff vigorously and with an impressive unanimity. In the summer of 1910 Sir Wilfrid Laurier, then prime minister, toured the prairie provinces. With skill and diplomacy he met hundreds of grain growers' delegations, and with infinite patience he listened for three months to scores of speeches and memorials urging tariff reductions. Sir Wilfrid was even more impressed with the growing political strength of the West, for the prairie provinces were winning a battle of statistics which threatened to alter the political balance of the dominion. The rapid settlement of the West posed for Canadian politicians the same problems that American politicians had faced many times.

[28] "The Agrarian Movement in Canada," *Quarterly Review* (London), 235:94 (January 1921).

[29] "The main force of criticism of the protective principle has come from British-born farmers." T. A. Crerar, "My Confession of Faith," *Grain Growers' Guide*, February 23, 1921.

The increase in prairie representation in the House of Commons from twenty to twenty-seven in 1906 forecast so clearly the coming change that even he who ran could read. Politicians, who do not run but move cautiously, were determined to keep this mounting strength in check. They could not allow it to disturb the political systems they had so carefully constructed. It would perhaps require concessions to the West, but not so many as to alienate eastern alliances. Canadian expansion had now reached the point at which politicians in Ottawa were faced with the problems in the reconciliation of sectional, racial, and religious diversities that had plagued their counterparts in Washington for many years. When Sir Wilfrid misjudged the political balance in 1911, he joined a distinguished company of North American politicians who had made a similar mistake.

The most dramatic move in the fight against the tariff came in December 1910, when eight hundred and eleven delegates representing agricultural societies from every section of Canada converged on Ottawa to present their demands to the Liberal government. Three hundred delegates represented the grain growers of the West in this "Siege of Ottawa," as it was called. The resolutions presented to the House of Commons contained nothing new; they urged reciprocal free trade with the United States in natural products and agricultural implements, an extension of the British preferential tariff, and complete free trade between Canada and Britain within ten years. In return the farmers declared a willingness "to face direct taxation in such form as may be advisable to make up the revenue required under new tariff conditions."[30] The delegates also pressed for a Hudson Bay railway, government operation of terminal grain-marketing facilities, revisions in the banking and railway acts, more favorable cooperative legislation, and government operation of necessary plants to establish a "modern and up-to-date" chilled meat industry. The tariff was the key issue, however, and of the twenty speeches delivered before the House, nine were concerned directly with the tariff and several of the others referred to it.

The Siege of Ottawa was perhaps as important as the farmers then believed, but not in the way they thought. Many farmers were

[30] George F. Chipman, *The Siege of Ottawa,* 4.

sure that the delegation which crowded the House on December 16 had convinced Sir Wilfrid that he must act. Though the government was impressed with the unanimity of opinion among eastern and western farmers, negotiations were already under way with the American government to secure much of what the farmers were demanding.

The Siege did serve two very useful purposes, however. It furnished the farmers' cause with publicity in newspapers in every section of North America. Issued under the skillful direction of George F. Chipman, the editor of the *Guide*, press releases informed thousands of the basic causes for the uprising which was developing in the Canadian West. The Siege also served as a meeting place for all the prominent agrarian leaders of Canada. The experience of working and fraternizing with rural leaders from every part of the dominion impressed the farmers with the strength of their class, and stimulated intersectional understandings that later were invaluable to the farmers' movement.

One interesting feature of the agitation for free trade has thus far escaped much analysis. This was the fundamental opposition of the farmers to any further industrialization in Canada. Their ears were deaf to patriotic arguments that the tariff "served the larger interests of the Dominion," for they maintained the physiocratic doctrine that an industrial state was one of poverty, disease, and tyranny. Farmers had no great concern for the effect of free trade on Canadian industry. They hoped, in fact, that by reducing the tariff they would prevent further industrialization. The "Farmers' Platform" in 1910 made this clear: "The greatest misfortune which can befall any country is to have its people huddled together in great centres of population and . . . the present customs tariff has the tendency to encourage that condition." [31] The population trend away from the farm would be encouraged if a protective tariff continued to build up Canadian industry. No government could remain democratic, thought the farmers, if industrialization and urbanization continued to diminish the ranks of the "true democrats of the soil."

The agitation for tariff reform reached a climax in the election of 1911. The ostensible issue of the election was the reciprocity

[31] "The Farmers' Platform" in Chipman, *op. cit.*, 4.

agreement that the Canadian and American governments had framed after months of negotiations. Concurrent legislation rather than a commercial treaty was to implement the agreement, which would allow natural products to move across the border without duty and would establish lower but identical customs rates on agricultural implements and other enumerated manufactured articles.

The two governments reached this agreement in January 1911, but it was July before the American Congress finally acted and late September before the Canadian election occurred. During this long delay North Americans had ample opportunity to observe the fate of economic issues which become involved in party politics, and the danger in a democracy of settling important questions on extraneous issues.

Impelled by a strong desire to halt the rising cost of living in the United States, to protect the dwindling popularity of the Republican party, and to give expression to a strong personal faith in international cooperation, President Taft vigorously supported the agreement.[32] While defending reciprocity, however, the president used an inept figure of speech which was later to help defeat in Canada what he had worked so energetically to win in the United States.

"The Dominion has greatly prospered," he said. "It has an active, aggressive and intelligent people. They are coming to the parting of the ways. They must soon decide whether they are to regard themselves as isolated permanently from our markets by a perpetual wall or whether we are to be commercial friends."[33] The "parting of the ways" figure was innocent enough in its context, but torn from their text the words were full of sinister meaning to loyal Canadians, especially to the British-born.

The grain growers faced a losing political fight in Canada. The public arguments advanced in the United States for reciprocity furnished Canadians with arguments against it. The adoption of the proposal by the American Congress was clear evidence to many Canadians that because the measure was designed to benefit the republic, it must therefore work to the disadvantage of the

[32] Henry F. Pringle, *The Life and Times of William Howard Taft*, 2:582–602.
[33] Quoted in *ibid.*, 589.

dominion. The mercantilistic conception of trade was strong on both sides of the line, and many believed that international trade was always at the expense of one of the trading nations. Western farmers fought this idea and pointed out that no sacrifices were required of the Canadian manufacturers under the proposed agreement. Economic argument met deaf ears, however, for Canadian manufacturers feared what the agreement presaged in later tariff understandings between the two countries.[34]

The defeat of reciprocity and the arguments used to discredit it were foreshadowed on the very threshold of the farmers' stronghold when the Manitoba legislature defeated a resolution approving reciprocity in March 1911. After briefly considering the economic effects on the transportation, merchandising, and financial interests of the nation, the legislative foes of reciprocity smothered the resolution with a flood of oratory which pictured the disastrous effects of reciprocity on the unity of the empire, imperial preference, and the "old flag." [35] Here in microcosm was the election of 1911 six months before it was held.

Economic issues were almost lost sight of in the hysteria that accompanied the election. Western farmers grew increasingly bitter as they saw their economic arguments faltering before the emotional attacks of an "Unholy Alliance" of the "vested interests" with nationalism, imperial sentiment, and French-Canadian nationalism. The *Guide* raised its voice in a vain effort to point out how patriotic sentiments were being prostituted by high tariff interests to advance economic privilege at the expense of the people of Canada.

A frequent note of hysteria entered the editorial columns of the *Guide* as public opinion turned against the farmers. The farmers' journal repeatedly denied any annexationist sentiments and emphasized the loyalty of the grain growers to the empire. The paper thought it saw a conspiracy between the Chicago and Winnipeg grain dealers to depress wheat prices and thus discredit reciprocity.[36] Canadian manufacturers had written the tariff long enough, cried the *Guide*; it was time for the farmer and the con-

[34] Vernon C. Fowke, *Canadian Agricultural Policy, The Historical Pattern,* 268.

[35] *Grain Growers' Guide,* 3:31 (March 8, 1911); *Canadian Annual Review of Public Affairs,* 1911, 119.

[36] *Grain Growers' Guide,* 3:5 (March 8, 1911).

sumer to be consulted in the economic life of the nation.[37] A final plea for urban support, which may also have served as a warning of the debacle to come, was advanced when the *Guide* pointed out that many farmers would be busy with the harvest and would thus be at a serious disadvantage during the campaign.[38]

The election of 1911 has been thoroughly examined in other studies,[39] but it is important to point out that no amount of argument from the grain growers could have prevailed. By September annexation, political absorption, and economic dependence were the issues which determined the voting outside Quebec, where the naval issue was the deciding factor. Farmer-advocates of reciprocity vigorously denied that free trade in natural products would mean political union, but they were wasting their breath. President Taft had unwittingly given substance to the argument when he had admitted that Canada was at the "parting of the ways," and Champ Clark proved the existence of the conspiracy in a foolish Congressional speech in which he repeated the theme so dear to politicians seeking Irish votes: "I hope to see the day when the American flag will float over every square foot of the British North American possessions, clear to the North Pole."[40]

The cry of loyalty to flag and empire served the purposes of the farmers' enemies. The economic issues were so effectively evaded or exaggerated that the probable effects of reciprocity were slightly known. A recent student of the subject has concluded that no great change in prices would have resulted in either country. Moreover, though local regions might have suffered or prospered under the agreement, neither the Canadian nor the American producer would have gained any substantial advantage.[41]

When the sound and fury had subsided, three facts were clear.

[37] *Ibid.*, 3:5 (February 8, 1911).

[38] *Ibid.*, 3:5 (August 2, 1911).

[39] L. Ethan Ellis, *Reciprocity of 1911* is the most complete study. Fowke, *op. cit.*, 265–68; Hugh T. Keenleyside, *Canada and the United States; Canadian Annual Review of Public Affairs*, 1911; Charles C. Tansill, *Canadian-American Relations, 1875–1911*; Wood, *op. cit.*, 237–70.

[40] *Canadian Annual Review of Public Affairs*, 1911, 62. During the election Clark was accused of numerous wild assertions. One patriotic society flooded Canada with a pamphlet containing an alleged speech declaring that the "Democratic party means to annex Canada by force or guile." Clark's denials had no effect, and thousands of Canadians believed that both parties had plans for annexing the dominion. Canadian newspaper and public opinion are sampled in H. F. Angus, *Canada and Her Great Neighbor*, 178–90.

[41] Ellis, *op. cit.*, 184.

It was again apparent that the Canadian farmers, despite their numerical strength, were as yet unable to influence governmental policy when their measures did not coincide with the interests of eastern capital. It was equally clear that Canada had come of age. The dominion had developed a national spirit the equal of any in the New World. And it was evident that what Mr. Dooley had said of the United States was equally true in Canada: "Thim that the tariff looks after will look after the tariff."

The defeat of the Liberals in 1911 was a severe blow to the grain growers. The *Guide* acknowledged that "the people of Canada have spoken," but added bitterly, "the rest of Canada must bow to Ontario, the most powerful province politically, in Canada."[42] The defeat of reciprocity forced the grain growers to look elsewhere for relief and thus stimulated demands for direct legislation and for a farmers' political movement. The agitation for lower freight rates also received additional support as a measure to offset the disadvantages of the protective tariff. The most interesting reaction, however, was the fluctuation in wheat prices which followed the news of the election. Quotations jumped five cents in Minneapolis and slumped two cents in Winnipeg.

The "Made in Canada" campaign which followed the victory of protection in 1911 was coolly received by western farmers. The *Guide* ridiculed the special trains that toured Canada with exhibits advertising the movement. The farmers' journal peevishly pointed out that the "Made in Canada" buttons, which enthusiastic protectionists displayed on their coat lapels, were manufactured in New Jersey. The official organ of the Canadian Manufacturers' Association, *Industrial Canada,* replied in kind by ridiculing the *Guide* for importing "free trade" books from the United States. To which the *Guide* replied: "Our policy is to buy where we can get the best value for our money."[43]

A favorite cartoon of the *Guide* during this clash with the manufacturers' association was to show a protectionist lecturing on the virtues of a high tariff with his apparel labeled "Scotch Tweeds," "foreign necktie," "hat made in England," "drinking glass made in Belgium," and "American made shoes and watch."

[42] *Grain Growers' Guide,* 4:5 (September 27, 1911).
[43] Quoted in *ibid.,* 4:6 (June 12, 1912).

The agitation for free trade continued despite the setback of 1911. In December 1913 the members of the Canadian Council of Agriculture presented their arguments for tariff reductions to the Borden ministry. This "Second Siege of Ottawa," however, was less successful than the first, and little encouragement was given the farmers.[44]

Hope for relief from the tariff burden came from an unexpected quarter. The American election of 1912 swept the traditionally low-tariff Democratic party into power. The Underwood Tariff Act, which followed in 1913, reduced the duty on wheat shipped into the republic from twenty-five cents to ten cents a bushel and offered to place wheat, wheat products, and flour on the free list if the dominion would place similar products on its free list. The actual value of this development proved to be slight, however, for the Borden government rejected the American proposal. In any case the action of the American government came too late to effect any substantial aid to the Canadian farmer, for the war halted the decline in wheat prices and provided a temporary demand in Europe that removed any necessity for entry into the American market.

The rising political strength of the prairie farmers did secure certain concessions from the Conservative government. The Canada Grain Act of 1912 continued the movement toward rigid regulation of the grain trade, and two further acts granted substantial financial aid to the provinces for agriculture. But these measures did little to increase the net return to the farmer for his products, and they were considered slight compensation for the rejection of reciprocity.[45]

The potential political strength of the western farmers was now too powerful to be satisfied with concessions, and the conviction was growing on the prairies that the time was ripe for the farmers to enter politics on their own initiative. Agitation for a farmers' political party grew rapidly, for conditions in western Canada favored a break with established practices.

The prairie provinces were youthful in outlook and possessed few political traditions to bind them to the established parties.

[44] *Ibid.*, 6:7 (December 24, 1913).
[45] Fowke, *op. cit.*, 246–47.

Moreover, settlers who had poured into the West from Britain and the United States had no sentimental ties with the Canadian Liberal or Conservative parties. Even settlers from eastern Canada seemed infected with political restlessness and under the impact of prairie conditions often lost their devotion to the party of their fathers.[46]

Faith in the existing parties was at low ebb on the prairies. Western farmers were convinced that neither party was concerned with the problems of the grain growers. They resented the Liberal campaign strategy which adopted the negative attitude, "if the Liberal tariff was unsatisfactory . . . a Conservative tariff would be doubly so."[47] Many farmers believed that a more positive approach to dominion politics could be made on their behalf. They felt that they were unrepresented in Ottawa, and complained that their cause was but slightly understood by the lawyers, professional men, and merchants who sat in Parliament.[48] Why, they asked, should Canadian laws be written by men who had no understanding of the problems of agriculture?

The revolt against "partyism" is dealt with elsewhere,[49] but it might be pointed out here that legislative bodies were often looked upon as the "tools of the interests." Party platforms were cynical catchalls, and "never more than something to fool the people."[50] Scandals, corruption, and party slush funds were cited as evidence of the unfitness of the regular parties to represent the farmers. The *Guide* wondered in 1914 if the parties "would be as solicitous for the welfare of the farmers as they now are for the manufacturers" if every one of its fifty thousand subscribers contributed ten dollars to a party slush fund.[51]

The reciprocity election of 1911 greatly weakened the strength of the traditional parties on the prairies. Many grain growers were convinced that neither party would give the farmer adequate representation, and independent farmer candidates sought election in

[46] John Nelson, *The Canadian Prairies: Their Problems and Policies*, 57.

[47] John W. Dafoe, *Clifford Sifton in Relation to His Times*, 343.

[48] For a discussion of this attitude see C. W. Peterson, *Wake Up, Canada! Reflections on Vital National Issues*, 34.

[49] See Chapter IV.

[50] *Grain Growers' Guide*, 3:6 (February 1, 1911); *ibid.*, 3:39 (April 26, 1911); letter of F. Kirkham; *ibid.*, 3:5 (May 10, 1911).

[51] *Ibid.*, 7:5 (April 8, 1914).

several prairie constituencies.[52] The directors of the Manitoba Grain Growers' Association were instructed by the annual convention to place before every candidate a copy of the Farmers' Platform and secure his commitment.[53] In Alberta a Farmers' Political Association was formed in one constituency with Rice Sheppard as executive secretary. The Saskatchewan Grain Growers' Association, however, rejected numerous pleas to organize a farmers' party, and reaffirmed its nonpolitical character. The United Farmers of Alberta also defeated a proposal to enter a farmer candidate in every constituency.[54] These decisions convinced many grain growers that an independent movement must be organized.

The crushing defeat suffered by the farmers in 1911 converted many who had been lukewarm in their support of political action. The *Guide* revealed the future hopes and plans of George Chipman, John Kennedy, and other prominent leaders in its reaction to the reciprocity defeat. "What is needed in Canada is a radical party with the courage of its convictions," declared Chipman in one editorial.[55] Subsequent editorials revealed that plans were under way for such a move. R. C. Henders, then president of the Canadian Council of Agriculture and the Manitoba Grain Growers' Association, spoke with authority when he denounced the "curse of partyism" and called for a new party of "the stalwart, intelligent yeomanry" combined with "the robust and independent labor party." [56] The Reverend S. G. Bland, the prominent reformer and Winnipeg clergyman, warned the S.G.G.A. that they "would never get the right men into politics, and never get the rascally men out, until we have a new political party." [57]

These were straws in the wind, and reveal the strength of the feeling that was sweeping the West. The demand for political action was so strong in Manitoba that one prominent grain grower was promised political reward if he would support the agitation for an independent party. When he refused to join the movement, his re-election as vice president of the M.G.G.A. was rejected.[58]

[52] *Ibid.*, 4:5 (August 30, 1911). [53] *Ibid.*, 3:5 (July 19, 1911).

[54] *Ibid.*, 4:5 (August 2, 1911); *ibid.*, 3:17 (February 1, 1911).

[55] *Ibid.*, 4:5 (September 27, 1911).

[56] *Ibid.*, 7:7 (January 14, 1914), presidential address before the M.G.G.A.

[57] *Ibid.*, 7:13 (March 18, 1914).

[58] The author has this information direct from the individual involved, who wishes to remain anonymous.

After the 1911 election the *Guide*'s columns were filled with elaborate plans for political action from scores of prairie farmers. The future course of the farmers' party was presented with amazing prescience by John Kennedy in a series of letters that reveal the political plans then under discussion by the staff of the *Grain Growers' Guide.*[59] Their real purpose was as much to secure farmer reaction as to enlighten. Kennedy suggested that the farmers call themselves Progressives and avoid "anything that savoured of a new party." The Progressives would be a "movement" directed by local committees in each constituency. The elected representatives in Ottawa would do nothing more "than vote the desire of their constituents" as revealed in "post-card referendums." He further predicted that such a move would create a "body of men in Parliament who will hold a balance of power for good legislation."[60] Kennedy prophesied accurately, for in 1921 these plans were realized.

Opposition to independent political action remained strong, however, despite the growing sentiment for a farmers' party. Enemies of political action marshaled their arguments skillfully and prevented any official entry into politics in the prewar years. They argued with the force of logic that the formation of a third party would be disruptive to the farmers' movement by reviving old party loyalties. Many feared that an emphasis on politics would endanger the cooperatives and might easily lead to a narrow class party.[61] Others pointed out the impossibility of securing united action from the farmers on clear-cut issues such as reciprocity and asked how the farmers could maintain any degree of unity "when a multiplicity of issues were involved?"[62] A further objection was the difficulty of financing a new party. Many feared that farmers would never part with sufficient cash to get a political organization functioning.

But the most effective argument against political action was the appeal to history. Most farmers on the prairies could remember the fate of the Grange in politics, and many were acquainted with

[59] This has been confirmed to the author by a journalist who was then with the *Grain Growers' Guide.*

[60] *Grain Growers' Guide,* 7:9 (March 4, 1914).

[61] James Morton, "Farmers in Politics," *Grain Growers' Guide,* 7:13 (August 5, 1914).

[62] Letter of W. S. Henry in *Grain Growers' Guide,* 6:15 (August 13, 1913).

the story of the farmer in politics in the United States. The conviction was widespread that any future attempt by the farmers in politics would be equally fatal to the grain growers' movement. Many Saskatchewan farmers, moreover, were convinced that political action was unnecessary since the provincial Liberal government was a farmers' party in all but name. Officials of the S.G.G.A., therefore, resisted the pressure to enter politics despite the bitter criticism that political inaction inspired.[63]

No amount of argument, though, could halt the growing sentiment that political action alone could save the western farmer. Some looked to direct legislation as an alternative, but the majority remained unconvinced that there was any alternative. An increasing number of grain growers' locals endorsed political action, and independent farmer-candidates entered provincial elections. By the time Canada was plunged into war in 1914, it was apparent that the mounting pressure was not to be denied. Only the return of peace, however, was to reveal what fate awaited this latest political crusade by the farmers.

Thus by 1914 the farmers had completed the cycle and were again advocating political action. Political revolt in the hard spring wheat belt broke out in North Dakota in 1916, but was delayed in western Canada by the war. When the uprising did hit the prairie provinces, two decades of indoctrination had disciplined the grain growers to act with unity. On top of this, twenty years of physiocratic education had convinced the farmers that the fate of millions rested on their actions. Economic justice and political democracy would survive in North America only through their efforts. Alberta farmers voiced the faith and hope of farmers of the entire Northwest when they sang:

Then come our friends, good and true
With good of all the world in view;
The die is cast, it's up to you
Organize! O Organize!

[63] *Ibid.*, 6:10 (April 9, 1913).

IV

The Prewar Crusade for Democracy

In the decade before 1914 a general democratic ferment on the Canadian prairies intensified agrarian discontent. During these years western Canadian farmers joined the crusade for greater democracy which was sweeping through North America. The protest against the concentration of wealth and political power in the hands of a capitalistic plutocracy thus spread into the latest frontier area on the continent.

The western Canadian farmer who protested against a high tariff, trusts and combines, and "money power" in 1911 did so in the best Jeffersonian tradition. His protests were rooted in the same soil of Lockean thought and evangelical Protestantism and sprang from the same grievances that had produced the Grange, the Farmers' Alliance, and Populism in the United States.[1] His crusade coincided with and sought the same fundamental objectives as the Farmers' Union, the Society of Equity, Robert M. La Follette's "Progressivism," and Woodrow Wilson's "New Freedom." This was no accident. The impact of monopolistic consolidation of Canadian industry hit the prairie farmer with such force during these years that in self-defense he turned to reforms similar to those advocated by American muckrakers and reformers in their "quest for social justice."[2]

North American farmers had long waged a losing fight to preserve agrarian democracy against the onslaughts of plutocracy and "money power." Jacksonian democracy, the Grange, and Populism were monumental efforts to halt the movement toward concentration. Canadian reformers shared with these the Jeffersonian faith in the virtue of the people and in the perfectibility of human institutions. They possessed a similar evangelistic fervor

[1] Chester McArthur Destler, "Western Radicalism, 1865–1901: Concepts and Origins," *Mississippi Valley Historical Review*, 31:335–68 (December 1944).
[2] Harold U. Faulkner, *The Quest for Social Justice, 1898–1914*.

for reform, for they too were in moral revolt against the corruption and sophistication of the system which threatened to overwhelm them.[3]

The "crusade against the New Feudalism," as the struggle against monopolism was called,[4] was led by the *Grain Growers' Guide*. Through editorials, feature articles, cartoons, and selected letters to the editor, this farmers' journal battled the "special interests" and exposed the forces which threatened to undermine democracy in Canada.

From the beginning, the *Guide* recognized that the crusade was continental in scope, for economic developments in the dominion had paralleled those in the republic. "The condition of affairs today in Canada and the United States drives home the striking truth that 'government for the people, of the people, and by the people' does not exist in the New World. The United States though technically a democracy, has in reality been a plutocracy. The wealthy classes in the United States have controlled legislation. In Canada . . . the same situation has existed."[5]

Canadian reformers were determined to prevent in their country the concentration of wealth and the growth of monopolies they had witnessed in the United States. One Saskatchewan farmer, who had come from New Jersey, phrased the question which many western Canadians were asking, when he wrote, "Can Canada afford to go through the same experience the States has—build up combinations and monopolies at the expense of the plain people, the laboring producing classes?"[6] The *Guide* was convinced that the only thing that could prevent hopeless misery for the poorer classes was "a revolution that will shake Canada to its very foundations."[7] This peaceful revolution would be accomplished through the legal exercise of the franchise, but it would constitute a thorough revision of Canadian economic and political life.

The conditions which impelled prairie farmers to radicalism were similar to those which had produced the agrarian revolt in

[3] W. L. Morton, "Direct Legislation and the Origins of the Progressive Movement," *Canadian Historical Review*, 25:279 (September 1944).

[4] This was adopted from Edward Porritt's influential volume, *The Revolt in Canada against the New Feudalism* (London, 1911).

[5] *Grain Growers' Guide*, 4:5 (November 1, 1911).

[6] Letter of Benjamin B. Hursh, Macoun, Saskatchewan, dated May 31, 1911, *ibid.*

[7] *Ibid.*, 3:5 (September 28, 1910).

the United States; and they inspired the same complaints: the complaints of a debtor economy. Insufficient capital to develop agricultural potentials, exorbitant interest rates, the control of farm prices by the manipulation of middlemen, and the protectionist policy of a government corrupted by "money power" were grievances on both sides of the forty-ninth parallel. Leaders in the Canadian reform movements were aware of the economic origins of their discontent, and they firmly believed that abuses in the economic system were responsible for the political and social inequalities which they saw in North America.[8]

The pioneer psychology was still strong on the Canadian prairies, and contempt for "eastern ways" was joined with a hatred for eastern financial and political control. This bitterness of spirit caused many thoughtful observers to consider the schism as one of the dominion's major problems.[9] The anti-eastern sentiment was often the result of objective analysis, but was sometimes the expression of a regional paranoia. The familiar denunciations of Wall Street were echoed in Canada by equally bitter condemnations of "St. James' Street" and the bankers of Montreal, and for similar reasons.

The chief cry of the revolt in western Canada was for the revision of the tariff, a reform which most American farmers had abandoned in favor of efforts to gain other economic controls, such as currency regulation and trust busting. Canadian farmers, though, were also keenly aware of the growth of trusts and monopolies, and energetically opposed the trend. Who owns Canada? This was a question often asked and easily answered by the readers of the *Grain Growers' Guide*. Evidence that only forty-two men controlled one third of the wealth of Canada in 1913 was sufficient to prove to the westerners that the New Feudalism had imposed itself on Canada as ruthlessly as it had on the United States.[10] Prairie farmers were convinced that the Canadian steel, soap, cement, canners, and Grocers' Guild mergers were the evil fruit of

[8] J. A. Stevenson, "The Battle for Democracy in Canada," *Grain Growers' Guide*, November 2, 1910.

[9] Charles W. Peterson, *Wake Up, Canada; Reflections of Vital Issues*, 224; J. O. Miller, ed., *The New Era in Canada*, 133; speech of Mayor Fleming of Brandon, Manitoba, as reported by *Grain Growers' Guide*, February 1, 1911.

[10] "Who Owns Canada?" *Grain Growers' Guide*, 6:9 (June 25, 1913); Porritt, *op. cit.*, viii; Gustavus Myers, *History of Canadian Wealth*, i.

the same "frenzied finance" and "water wagon finance" that had cursed the American farmer and laborer. The conviction that most of these mergers were "nothing better than legalized conspiracies against the savings of the people" was an article of faith in the West.[11] The grain growers were fully alert to these developments and by 1914 were in open revolt against the New Feudalism.[12] They believed that they had launched "the most important popular political movement in Canada since the rebellion of 1837," a movement which would "reestablish the proper functions of representative institutions for the people of Canada and renovate the whole system of national life."[13]

The objectives of the grain growers' fight for democracy were the same as those of the American agrarian movements. Fundamentally, they hoped to strengthen capitalism by saving small enterprise from destruction. They sought no less a goal than to wrest possession of the government from the plutocracy, and use it for democratic rather than plutocratic ends.

The farmers of the prairie provinces were bitterly criticized by many eastern Canadians for following the lead of "Yankee republican ideas." It is now clear that most of this criticism was based on a misunderstanding of prairie conditions. Western Canadian farmers, in most cases, were not consciously advocating "American reforms," as was so often charged. They were trying desperately to find answers to problems which were North American in nature and continental in scope. It should not be surprising that they proposed solutions which paralleled those suggested or tried in the United States. Alberta, because of the great numbers of Americans who had migrated to that province, was most frequently accused of supporting "peculiar" and "radical American" proposals.

Oklahoma agrarians during these years were urging the same reforms that were supported by Alberta farmers. The radical agrarian program of the Farmers' Union and the radical agrarianism of William H. "Alfalfa Bill" Murray, the "Sage of Tishomingo," strikingly paralleled the western Canadian revolt. Murray's extemporaneous address in the opening sessions of the constitutional

[11] *Toronto Telegram*, quoted by Porritt, *op. cit.*, 60.
[12] Porritt, *op. cit.*, 227.
[13] Stevenson, *op. cit.*, November 2, 1910.

convention in 1906 could have been delivered before a grain growers' convention with few changes.[14] His denunciations of land monopoly, political power of the special interests, corporate greed, and political graft, fraud, and bribery would have been cheered as enthusiastically in Calgary as they were in Guthrie.

Explanations for this parallel development are not hard to find. Both regions were the latest settled, for Alberta was granted provincial status in 1905 and Oklahoma did not enter the American Union until two years later. Both regions were consistently debtor economies with insufficient monies to underwrite agricultural expansion, and both lacked stability in political party life. Though radical agrarians in each of these regions advocated similar reforms, there is no evidence that either group ever heard of the other, or found any inspiration in the other's work.

This was perhaps an extreme case. That there were American influences in Alberta and in the Canadian West is clearly true. But it was often the case that what were in fact parallel movements were called "American influences" by those blind to the larger concept of the common problems of North American agriculture.[15]

The protest movement on the prairies was nourished by three dynamic traditions of radicalism. The earlier agrarian movements in eastern Canada, the British cooperative movements, trade unions, and Socialism, and the agrarian movements in the United States furnished a background of experience and inspiration for western Canadian farmers.

Many western farmers had earlier been members of the Grange or Patrons of Industry in eastern Canada or had witnessed the growth of these two farmers' movements in Ontario. Though failures, these earlier movements prepared the way for the later grain growers' organizations. Farmers were indoctrinated in the basic necessity for group action, and men like E. A. Partridge and J. W. Scallion were trained for a greater work in the West.[16] W. L. Smith, for many years editor of the *Farmers' Sun* and intimately

[14] *Proceedings of the Constitutional Convention of the Proposed State of Oklahoma,* 15–25 (Guthrie, 1907).

[15] For an illustration of this confused thinking, see "The Agrarian Movement in Canada," *Quarterly Review* (London), 235:84–107 (January 1921).

[16] Patrons of Husbandry, *Journal of the Proceedings of the Dominion Grange,* 25th Annual, 1909.

acquainted with Canadian agrarian movements, was convinced that "the whole farm movement in the West may fairly be styled as the joint progeny of the Patron movement of Ontario and that like movement which swept across the border from the Western States."[17]

Western Canadian agrarianism was further enriched by the background in British cooperatives and trade unionism possessed by so many grain growers. Thousands of prairie farmers had been reared in an atmosphere of social ferment in Britain, and many had belonged to cooperatives or trade unions in the old country. They were well acquainted with the Newcastle Program of the early 1890's and followed with keen interest the revival of social reform under Asquith and Lloyd George. The *Guide* kept its readers informed of these developments and drew constant inspiration from the British labor movement. This background of trade unionism and cooperation furnished much of the program of the grain growers and supplied much of the prairie leadership. Many believed that radicalism on the prairies was the direct result of the British trade union experience,[18] and immigrants from the British Isles were struck with the similarity between the radicalism they had left in Europe and that they found in the New World.[19]

The influence of American radicalism was also present. The effects of geographical propinquity and immigration were easily recognized as responsible for this influence. Equally important, but less easily recognized, was the influence of American publicists and muckrakers. The wide circulation of American magazines and journals on the Canadian prairies brought into Canadian homes the crusade against political corruption and economic exploitation conducted by Lincoln Steffens, David Phillips, Charles Edward Russell, and others in *McClure's Magazine*, *Collier's Weekly*, *Cosmopolitan Magazine*, and *Pearson's Magazine*. American novelists were also influential in Canada. The language and vocabulary of the leaders of the United Farmers of Alberta was that of Hjalmer Hjorth Boyesen's *Mammon of Unrighteousness*.

[17] W. L. Smith, "When the Patrons Swept Ontario," *Grain Growers' Guide*, 19:6 (December 1, 1926).

[18] John Foster Fraser, *Canada As It Is*, 111, and Agnes C. Laut, *Canada at the Cross Roads*, cite illustrations of this opinion.

[19] Letter of John MacDonald, *Grain Growers' Guide*, April 19, 1911.

The social protests of Henry George's *Progress and Poverty*, Edward Bellamy's *Looking Backward*, Frank Norris' *The Octopus* and *The Pit*, and Upton Sinclair's *The Jungle* had a profound influence in Canada as in all English-speaking countries. Canadian reformers found much of their evidence against society in these books and from them they secured their vocabulary of social protest.

The cross-fertilization of ideas provided by this diversity of influences created a breadth of understanding which was sometimes lacking among agrarian reformers in the American West. The Canadian reform movement was enriched and broadened by the confluence of these three streams of radicalism.

The growing influence of the "social gospel" also contributed to the agrarian crusade in western Canada. The strong moral tenor of the grain growers' organizations gave the crusade a religious fervor which appealed to thousands of devout prairie farmers. Poverty, ignorance, and hatred could be eradicated from the world "if the Heavenly Spirit—this spirit of Brotherhood—could be instilled as the working creed of the world."[20] The ruling spirit of the "God of Mammon" was clearly seen in the actions of the Liberal and Conservative parties, which were thus unfit to be the instruments of public will.

This moral revolt against political corruption and economic tyranny was strongly supported by a small but able group of clergymen who were motivated by a growing recognition of the social implications of the Christian gospel. These clergymen were clearly under the influence of European and Eastern theologians. They were also greatly inspired by the writings of the American novelist Winston Churchill. Churchill, the most representative of the Spirit of Progressivism,[21] emphasized the immorality of contemporary political life and the abusive control exercised over the church by accumulated wealth. His novel *The Inside of the Cup* became the textbook of the "social gospel" reformers and was widely read and discussed in western Canada.

The real leaders of the social gospel movement in western Canada were the Reverend Salem G. Bland, a professor at Wesley

[20] Letter of Bert Huffman, *ibid.*, 4:8 (July 3, 1912).
[21] V. L. Parrington, *The Beginnings of Critical Realism*, 348.

College in Winnipeg,[22] and the Reverend J. S. Woodsworth, then secretary of the Social Welfare League. Through sermons, lectures, and frequent articles, these two able clergymen expressed their conviction that the grain growers, trade unions, and cooperative movements were "applied Christianity." Moreover, it was the Christian duty of every farmer to support these movements in the "real and bitter war between capital and labor." [23] The social gospel was explained to prairie farmers in weekly "Sermons for the Unsatisfied" by Woodsworth in the *Guide*. Both men threw their energies into all phases of reform through their genuine concern for the welfare of the common people.[24] As a result, both were bitterly condemned by the conservative press and clergy for their activities.

The moral tone of the crusade excited the attention and support of many prairie clergymen. R. C. Henders, long-time president of the Manitoba Grain Growers' Association, who later fell from grace and was expelled from the movement, began his career as a clergyman, as did the more successful William Irvine in Alberta. Henry Wise Wood and Percival Baker, who were soon to emerge as leaders of the United Farmers of Alberta, were trained in ministerial colleges in the United States, and both did lay preaching in the Christian Church.[25]

To the work of these men must be added that of scores of ministers in village and rural parishes who shared the hardships and aspirations of their farmer congregations. The paucity of trained leadership in many communities often placed the pastor in the role of spokesman for the farmers. When the farmers later entered politics, more than one of these western clergymen dreamed of becoming the Moses who would lead his people to the promised land in Ottawa.

The rising labor movement in the growing cities of the West

[22] The union of the Methodist and Presbyterian churches in Canada brought the merger of Wesley College and the Presbyterian college, Manitoba College, to form United College.

[23] Sermon by the Reverend S. G. Bland in *Grain Growers' Guide*, 8:15 (March 31, 1915).

[24] Both clergymen were frequent speakers on behalf of direct legislation, prohibition, direct tax on land values, trade unionism, and pacifism.

[25] Wood in Culver-Stockton College, Canton, Missouri, and Baker in the College of the Bible, Lexington, Kentucky. Both institutions are affiliated with the Disciples of Christ denomination.

was also in league with the farmers during these years. The grain growers' movement and trade unionism found a community of interest in the struggle against the plutocracy until the extreme radicalism and violence of the famous Winnipeg strike in 1919 shocked many farmers into opposition. Until that time, prominent labor leaders cooperated with the grain growers in support of direct legislation, the single tax, a graduated income tax, and other reforms. F. J. Dixon of Winnipeg and William Irvine of Calgary were typical of this unity of leadership and purpose in that they promoted these reforms in both movements.

Officials of the grain growers' organizations were friendly to the cooperative action of the two groups and were supported by large numbers of individual farmers who realized the need for unity. Many grain growers thought the laboring classes, if "properly enlightened," were valuable allies.[26] The U.F.A., for example, in its annual convention in 1911 unanimously adopted a resolution creating a committee to confer with organized labor "to ascertain if there be any neutral ground on which to work out our common interest."[27] Suspicion and distrust on both sides, however, caused many pleas for farmer-labor cooperation to fall on deaf ears. After 1919 western farmers often feared "Bolshevik" labor leaders as much as protectionist industrialists, and hopes for political cooperation in a crusade against the "vested interests" collapsed.

Radical westerners condemned the New Feudalism as a menace to Canadian democracy. The growth of money power was accompanied, they insisted, by a corruption of government, a venal press, poverty, and economic and political tyranny. Western farmers had little confidence in their Liberal representatives in Ottawa, who, for the sake of power, had "betrayed" the farmers and had deserted their low tariff stand of 1894.

The farmers believed they stood as the only organized group against the iniquities of the tariff, government subsidies, and monopolies. They were convinced that as a class they carried the burdens of the economic system without sharing its advantages. They maintained the physiocratic doctrine that they created the

[26] *Grain Growers' Guide*, 4:8 (June 19, 1912); *ibid.*, 7:15 (June 3, 1914).
[27] *Ibid.*, 3:17 (February 1, 1911).

wealth of the nation, and that the prosperity of all other classes rested on their work. The *Guide* often published cartoons from the clever pen of Arch Dale illustrating the dependence of businessmen, laborers, mechanics, clergymen, lawyers, and doctors upon the work of the farmer.

The fervor with which prairie farmers took up their battle against the New Feudalism is demonstrated in campaign songs which were popular during this period.

The Day of Right

(TUNE: "THE BATTLE HYMN OF THE REPUBLIC")

The farmers of the prairie lands are massing in their might,
Exulting in a Principle, a Cause for which they fight;
The sacred cause of Justice, the establishment of Right,
And Equal Rights to all.

REFRAIN:
Oh! 'Tis time to get together;
Join and help us get together;
We have vowed to stand together,
For the day of Peace and Right.

The farmers of the prairie lands have right upon their side,
Their platform is the people's—democratic, nation-wide;
Their cause, the ancient cause for which brave-hearted men have died—
Of Equal Rights to all.

The farmers of the prairie lands know well the foe they fight,
The profiteers of Privilege, full armed with legal right;
Against that giant bluff we need to solidly unite,
For Equal Rights to all.

The farmers of the prairie lands today extend a hand
To town and country, East and West, where men for freedom stand;
Their "fiery cross" flames out today and every field is manned
For Equal Rights to all.

The farmers of the prairie lands, their wives and kith and kin,
Link up today true-hearted all to help the fight to win;
Assured that for our Canada a new day will begin,
With Equal Rights to all.

Marching to Victory

(TUNE: "MARCHING THRO' GEORGIA")

Bring the good old bugle boys; we'll sing another song,
Sing it with a spirit that will help the world along,
Sing it as we now can sing, two hundred thousand strong,
As we're marching to vict'ry.

REFRAIN:

Hurrah! Hurrah! We fight for Equity;
Hurrah! Hurrah! A world democracy;
Fight, from exploitation every brother to make free;
As we're marching to vict'ry.

We must carry on the fight, no matter when it ends,
Labor men and soldiers too, will join us as their friends;
On our solid unity the good of all depends,
As we're marching to vict'ry.

Fighting for democracy, with ballots in our gun,
Tyrants and the plutocrats—we'll beat them as the Hun;
All the hordes of privilege will soon be on the run;
For we're marching to vict'ry.

The farmers' revolt against "partyism" was based on the conviction that the traditional parties were so corrupt and unrepresentative of the people that democracy itself was at stake in Canada. Regular parties were associated with graft, booze, a subsidized press, bossism, protection, debt, bribes, greed, and political corruption.[28] One Canadian muckraker believed that the "root of the evil lies largely in our economic life. It corrupts our political system, our political system corrupts and degrades the public administration, and the corroding influence extends to the social system and business life until the disease permeates the whole community."[29]

Self-government was a myth, thought others, for Canada was oppressed by the rule of industrial and political monopolists.[30] Many prairie farmers applauded Edward Porritt's dictum that "statesmanship and public service cannot exist side by side with

[28] *Grain Growers' Guide*, 8:6 (May 12, 1915).
[29] Stevenson, *op. cit.*, November 2, 1910.
[30] Speech of President R. C. Henders of the Manitoba Grain Growers' Association before the annual convention of 1912, *Grain Growers' Guide*, February 7, 1912.

the New Feudalism."[31] The "interests" had no use for statesmen when their purposes were best carried out by subservient "peanut politicians."

Agrarian discontent reached such proportions by 1911 that Laurier altered the course of Liberal policy, and his government went down to defeat as a result of his efforts to placate angered westerners.

Western radicals were discouraged by the disaster of 1911, but they continued the fight in the belief that "democracy must ultimately triumph." The defeat of 1911 did not end the agitation for reform, though emphasis on the tariff lessened in favor of demands for direct legislation and for the formation of a third party. By 1914 the conviction was widespread in the prairie provinces that it would be futile to pour the new wine of reform into the old wine-skins of partyism, but by that time a newer and greater menace threatened the democracy of Canada.

Radical agrarians of these years of ferment saw the Canadian Manufacturers' Association as the chief enemy of democracy and political righteousness in Canada. The "battle of the people versus privilege" was chiefly fought against this citadel of "reaction and corruption."[32] The official journal of the C.M.A., *Industrial Canada,* was carefully watched for "democratic heresies," and was widely known as the "journal of the New Feudalism." Injudicious and extravagant statements by C.M.A. officials were quoted by agrarian radicals to prove the sinister purposes of the organization.

Western farmers were outraged in 1910 when the secretary of the association boasted that "the reorganized C.M.A. is like a young giant ignorant of its own power. By the exercise of these powers it could, if it chose, bring several millions of people to the verge of starvation, or paralyze the industry of the whole Dominion."[33] The *Guide* considered this to be the "most outstanding challenge . . . ever hurled at the public in many years. To think that 2,500 men in Canada claim that they are able to make the grass grow in the streets of this whole great country is something for every sane man to consider."[34]

[31] Porritt, *op. cit.,* 181.

[32] *Grain Growers' Guide,* February 7, 1912.

[33] Speech of G. M. Murray, Winnipeg, February 9, 1910, quoted by Porritt, *op. cit.,* 108.

[34] *Grain Growers' Guide,* 2:5 (February 9, 1910).

Western agrarians grew increasingly cynical of the optimistic prophecies and gratuitous advice offered by the plutocracy. The pious platitude that if "Western farmers should look to their ultimate welfare . . . they would see that a uniform national development is worth making sacrifices for"[35] was met with derision as a typical example of the free advice intended to hoodwink the farmer. The *Guide* warned its readers that most of the advice offered by the conservative eastern press was pure "hokum." "Always bear in mind," cautioned the editor, "that the good old patriotic slogan of the C.M.A., 'Canada for the Canadians,' means Canada for 2,500 Canadians."[36] The feud between the *Grain Growers' Guide* and *Industrial Canada* was marked by the great bitterness which so often characterized the resentment felt in western Canada toward the industrialists as a "pressure group."[37]

Other journals also fell under the condemnation of the *Guide*. The *Toronto Saturday Night* was frequently criticized for its "ignorance of prairie conditions" and its "sophistication." The *Guide* often denounced the undemocratic editorial policy of the *Canadian Century*, which it considered to be the mouthpiece of Sir Max Aitken and the notorious cement trust. When the *Century* suspended publication in 1912, the *Guide* published a satirical obituary entitled "One Enemy Fallen," and reported that "there were few mourners."[38]

This bitterness of spirit grew out of the conviction that freedom of the press no longer existed under the New Feudalism. "Freedom of the press in Canada is a myth," lamented the editor of the *Guide*, as he surveyed the newspapers and journals of the dominion. "Practically every politician of any size in Canada," he believed, "has a newspaper in his hip pocket." This vicious control over the press was certainly "one of the reasons why Special Privilege and unscrupulous politicians flourish in Canada." The venality of the press was presented with convincing evidence to the prairie reader, and the social implications of the decline of a free press were widely understood. The *Guide* was convinced that if the

[35] *Industrial Canada*, September 1910.
[36] *Grain Growers' Guide*, 2:5 (February 23, 1910).
[37] S. D. Clark, "The Canadian Manufacturers Association," *Canadian Journal of Economics and Political Science*, 4:505–23 (November 1938).
[38] *Grain Growers' Guide*, 4:6 (January 24, 1912).

control of the newspapers could be placed in the hands of the journalists "there would be a revolution in Canada inside of five years, and Special Privilege would be wiped out."[39]

The *Guide* dramatically announced to its readers in 1912 that the farmers' journal itself was under pressure from advertisers to alter its editorial policy or lose their support. Advertisers were objecting that the *Guide* "stirred up ill-feeling between the farmers and the manufacturers instead of endeavoring to get them to work together . . ." The issue before the paper was clear-cut. "Shall the *Guide* be compelled to betray the farmers of the West, or shall it remain free to tell the truth?"[40] The answer was obvious, and the new subscriptions which poured into the *Guide* office were a vote of confidence from angry grain growers.

The New Feudalism was also held responsible for the dangerous schism between East and West. Western agrarians firmly believed that the concentration of wealth and political power in the East had created a government whose only concern was the industrialist. Resentment against the East, as a result, grew so vehement that the *Guide* reluctantly admitted the presence of a secessionist sentiment in the prairie provinces. Threats to withdraw from the Canadian confederation and to establish a new British dominion in the West were deplored by the *Guide*, which believed that if the common people of East and West would "get closer together rather than farther apart, they would quickly shake off the shackles they are forced to carry," and secessionist agitation would soon die out.[41]

Though the *Guide* discouraged secessionist talk, it did believe that a "few more experiences like that of 1911, with partial failure of crops, and a total failure to provide a market for what we did have will increase discontent."[42] Grievances against eastern Canada could be removed, however, by the defeat of eastern "money power." Such a victory would open the way to an "uncorrupted" representation in Ottawa, and thus bring the legislation necessary for the welfare of prairie agriculture.

[39] *Ibid.*, 4:5 (September 20, 1911); *ibid.*, 3:5 (October 26, 1910); *ibid.*, 4:5 (September 20, 1911); *ibid.*, 3:5 (October 12, 1910).

[40] *Ibid.*, 4:5 (April 17, 1912).

[41] Memorial presented to Robert Borden by farmers at Yellowgrass, Saskatchewan, *Grain Growers' Guide*, 3:5 (July 5, 1911).

[42] *Ibid.*, 4:22 (June 26, 1912).

Impelled by a strong Jeffersonian faith that direct political action could perfect social institutions, western agrarians launched the crusade to create a democratic society in Canada. The power of the New Feudalism would be crushed, and agrarian plans for peace, democracy, and equality would rule throughout the dominion. The economic reforms proposed to bring all this into reality in Canada closely paralleled those which were advocated south of the border to democratize American society. "Special Privilege" in Canada, it was maintained, could be removed by curbing monopolies, destroying the protective tariff, discontinuing governmental bounties and subsidies to industry, and altering the tax structure to a single tax basis.

The most spectacular effort to defeat the New Feudalism was the fight to lower the tariff, and in this the western farmers met defeat and discouragement. Agrarian efforts to secure free trade, reciprocity, and a lower tariff have been pretty well explored,[43] but it is worthy of notice here that agrarian economists in western Canada advocated a graduated income tax to replace revenues lost to the government should their program of tariff reduction be adopted. This tax was strongly supported as less oppressive than the tariff, and its function in reducing the strength of money power was clearly understood. After the Canadian people rejected tariff reform in 1911, the graduated income tax remained as a plank in the agrarian platform; but the emphasis shifted to the social function the tax would perform in reducing accumulated wealth as "a stepping stone to the ultimate goal" of a more equitable distribution of wealth in Canada.[44]

Anti-monopolism was a key issue to those who understood the relationship of monopolies to the economic disabilities under which the farmers labored. Trust busting was advocated in Canada as in the United States, and with about the same measure of success. The dominion parliament acted to curb trusts in Mackenzie King's Combine Investigations Act of 1910, which defined a combine as "an association whose activities were detrimental to the

[43] L. Ethan Ellis, *Reciprocity of 1911*; V. C. Fowke, *Canadian Agricultural Policy, the Historical Pattern*, 265–68; *Canadian Annual Review of Public Affairs*, 1911; Louis A. Wood, *A History of the Farmers' Movements in Canada*, 237–70.

[44] *Grain Growers' Guide*, 7:5 (July 8, 1914).

public good."[45] The *Guide* was completely dissatisfied with this bill, for it sought "to prevent combines without removing the cause. It is rather like placing a nice tempting bone before a dog, and then training him not to touch it under pain of severe punishment. Trouble would be avoided by removing the bone."[46] The *Guide's* objections proved to be valid, for only one combine was ever investigated under the act and mergers continued apace.

Efforts to secure tax reform in the prairie provinces were more successful. The prewar years were marked by a strong and well-organized movement to adopt Henry George's single tax. In its earliest issues the *Guide* promoted the single tax scheme, and conducted a highly effective campaign of indoctrination. Through the financial support of the American millionaire, Joseph Fels, the *Guide* was able to offer its readers a "very fine library of ten volumes" of the works of Henry George.[47] Articles, letters, and selected quotations from the Chicago soap king frequently appeared in the *Guide* in support of the single tax doctrine.

By 1914 the three prairie provinces had adopted certain aspects of the "single tax" plan. In 1913 Alberta abolished the flat tax per acre on farm lands in favor of an unearned increment tax, a wild lands tax, and a lumber acreage tax. Saskatchewan followed this lead the next year with similar legislation, including a stiff surtax on uncultivated lands and a rural municipalities' hail and weed tax. Manitoba heavily taxed unoccupied lands and exempted from taxation all farm produce, agricultural implements, and farmers' personal property. In all three provinces agrarian legislators exempted land improvements from taxation, and land speculation was discouraged by tax laws that favored cultivated and occupied lands.

Tax legislation in the prairie provinces was one of the most important, though infrequently mentioned, victories of the agrarian reformers in North America. It was the envy of American farmers.[48]

[45] Mary Q. Innis, *An Economic History of Canada*, 288.
[46] *Grain Growers' Guide*, April 27, 1910, quoted by Porritt, *op. cit.*, 60.
[47] *Grain Growers' Guide*, 3:5 (May 3, 1911).
[48] Frank Packard, "Tough on Speculators, How the Northwestern Provinces of Canada Have Lightened the Burdens of the Producing Farmer by Taxing Idle Land," *Nonpartisan Leader* (Fargo, N.D.), 2:3 (March 9, 1916).

Political democracy was to be achieved in Canada, thought radical westerners, primarily through the use of direct legislation: the initiative, referendum, and recall. Here again, North American reformers placed implicit faith in the efficacy of legislation to eliminate the evils which corrupted democracy in the early twentieth century. Demands for direct legislation became increasingly insistent on the prairies after the defeat of reciprocity in 1911. To many grain growers, it was the only alternative to a third party,[49] and to others it was a convenient vehicle for "pet reforms."

Sweeping claims for direct legislation and a naive faith in its effectiveness characterized the agitation for the reform. Direct legislation, it was claimed, would place political power in the hands of the people and thus end graft, injustice, corruption, patronage, and the misuse of political power by "corporation politicians." "Direct legislation will put a loaded gun into the hands of the people and the 'era of misrule' will end" was a typical exaggeration.[50]

The adoption of direct legislation by states in the republic was cited as an argument for its use in western Canada. The experience of Oregon was watched with particular interest, and grain growers were told in their official journal that Oregon "in a great measure had eradicated the evils with which most of us are familiar by giving the best that is in all of us a chance to express itself." Because of direct legislation, Oregon was declared to be the "most complete democracy in the world," and election days "really meant something" to the common people of that favored state.[51] Progress of the reform in the States was followed with keen interest, and prairie farmers were encouraged that even in Arkansas direct legislation had put "new spirit into the people . . . made them feel new responsibilities and inspired them to broader information and higher intelligence." As state after state accepted the reform, the *Guide* was moved to challenge its readers at a sensitive point: "Shall we lag behind our cousins to the South?"[52]

[49] The S.G.G.A. declared against a third party because legislation could be controlled by the initiative and the referendum (*Canadian Annual Review of Public Affairs*, 1913, 609).

[50] *Grain Growers' Guide*, 3:5 (February 1, 1911).

[51] Frederick C. Howe, "Oregon: The Most Complete Democracy in the World," *Grain Growers' Guide*, May 3, 1911; *ibid.*, 4:5 (September 20, 1911).

[52] *Ibid.*, 4:5 (May 8, 1912).

The agitation for direct legislation was spearheaded by the Direct Legislation League, which was affiliated with the Manitoba Grain Growers' Association, the Royal Templars of Temperance, the Winnipeg Trades and Labor Council, and the Manitoba League for the Taxation of Land Values. The league was heavily supported by the "inspired millionaire," Joseph Fels, who doubled the total amount of subscriptions in 1910 and 1911. Fels was widely acclaimed on the prairies as one "who prefers to distribute his wealth to movements for the abolition of social inequality," and his generosity was often relied on to provide financial support for a variety of reform schemes.[53]

The grain growers' movements enthusiastically supported direct legislation, and by 1911 all three prairie societies had officially endorsed the reform.[54] The United Farmers of Alberta decided that year that every officer and director of the organization should be pledged to work for such legislation.[55] The organized farmers believed that through direct legislation they would be able to secure the specific class legislation the farmers needed, and at the same time purify the political system.

The demand for direct legislation came under severe criticism as the possible source of much hasty, crude, and impractical legislation. Opponents in Canada were able to point out, moreover, that direct legislation was clearly unconstitutional, for sovereignty under the British system rested with parliament, not with the people.[56] Enemies of the reform also denounced it as an American innovation and fad, and rang the changes on the dangers to British institutions of a device so obviously inspired by American republicanism.

Premier Sir Rodmond Roblin of Manitoba led the emotional attack against direct legislation. Debates in the Manitoba legislative assembly were punctuated by caustic reminders that Liberal supporters of direct legislation were forgetting Canada was a British, not an American, country.[57] The legislature was so moved by the premier that it resolved that "the British form of responsible gov-

[53] Seymour J. Farmer, "Direct Legislation in Manitoba," *Grain Growers' Guide*, December 6, 1911; *ibid.*, 4:5 (November 1, 1911).

[54] W. L. Morton, *op. cit.*, 282.

[55] *Grain Growers' Guide*, February 1, 1911.

[56] Morton, *op. cit.*, 287.

[57] *Ibid.*, 286.

ernment, as enjoyed by the Province of Manitoba, is the best form of government in the world."[58] Earlier, the patriotic premier had denounced direct legislation as "degenerate Republicanism." When the Conservative government fell in 1915 as a result of the Parliament building scandal, the *Guide* was unkind enough to remind Sir Rodmond that the scandal could have been prevented by direct legislation, and that "it would have protected him from his own sins."[59]

The results of the agitation for direct legislation were meager, considering the number of organizations which joined the movement. Premier Sifton's government in Alberta put through a direct legislation bill of "questionable value" in 1913, and in the same year the Liberal government in Saskatchewan passed an act providing for direct legislation if the act itself were ratified by 30 percent of the electorate in a referendum. The act excluded financial legislation from public control, and the grain growers were so dissatisfied with it that they failed to sustain it in the referendum.[60]

The Manitoba legislature also passed a direct legislation act which excited popular interest as it moved up through the courts to the Judicial Committee of the Privy Council, where it was ruled *ultra vires* of the provincial legislature. The reaction of the supporters of direct legislation in the Canadian West to this decision was typical of North American democracy. "This may be the law, but it is not the fact," objected the *Manitoba Free Press*, and the *Guide* prophesied that "direct legislation is a development of democracy which their Lordships over in London cannot successfully stand in the way of."[61]

Agitation for direct legislation gradually died out as experience in the United States showed North Americans that it possessed no magic powers to purify politics. With the coming of the war, direct legislation was forgotten as the grain growers turned their energies to more pressing problems.

Direct legislation did not use up all the energy of agrarian re-

[58] *Canadian Annual Review of Public Affairs*, 1913, 551.

[59] *Grain Growers' Guide*, 8:5 (May 19, 1915).

[60] *Ibid.*, 7:5 (February 4, 1914). The *Guide* thought that 30 percent was "far too high," and explained the low vote was due to the fact that the referendum did not coincide with a regular election (6:6, December 3, 1913).

[61] *Manitoba Free Press*, December 21, 1916, quoted by Morton, *op. cit.*, 287; *Grain Growers' Guide*, 12:6 (July 9, 1919).

formers. They worked for nearly all the reforms advanced in North America before 1914 to broaden the benefits of democracy and protect it from its enemies.

To purge the cities of their political corruption and give their fellow citizens a "business-like" government, the grain growers often advocated the commission form of city government.[62] This new form of municipal administration had proved successful in several American cities, and many saw in the system an answer to similar problems in Canada.

To bring "common justice to wives, daughters and sisters," the farmers supported the agitation for woman suffrage, though not without misgivings. That they were less ardent for this reform than for others was clearly revealed when the *Guide* conducted a poll of public opinion in 1913. Woman suffrage was supported by a two to one margin, but this was far below the majorities given direct legislation, public ownership of telephone and telegraph facilities, and other vital economic reforms.[63]

By 1913, however, the grain growers in all three provinces had organized women's sections, and by 1916 the women were full-fledged members of the farmers' societies. This happy turn of events "helped greatly to save the United Farmers' movement from the usual fate of male movements," wrote William Irvine in later years. "The male mind, during the individualistic system of society, went to seed on commercialism; and the human values, such as character, the development of the aesthetic tastes, and the desire for the highest happiness, have been choked out."[64] Canadian reformers shared the illusions of American liberals that woman suffrage would smash corruption, purify politics, and place the life of the nation on a higher plane.

To protect minority groups the reformers frequently advocated the adoption of proportional representation. Though this reform was rarely adopted in the United States, it was more successful in western Canada. Several important communities in the prairie provinces conducted their elections under this principle as a result of this agitation. Radical westerners also joined the movement to reform the Canadian Senate and raised their voices in protest

[62] *Ibid.*, 4:5 (June 19, 1912).
[63] *Ibid.*, 6:7 (February 12, 1913).
[64] William Irvine, *The Farmers in Politics*, 124.

against gerrymandering, "saw-offs," and the conferring of British titles on Canadian politicians.[65]

Social democracy was an ambition which Canadian reformers cherished. They believed that only by giving women their rightful position in society, by defeating demon rum, and by preserving the peace of the world could democracy thrive. The temperance movement which swept North America during these years was as strong on the Canadian prairies as in any rural community on the Kansas plains. The constant support of the *Guide* for the "Banish the Bar" agitation and the enthusiastic condemnation of the liquor traffic by the grain growers' associations bore fruit during the war when the prairie provinces joined the continental effort to alter the habits of a people by legislative enactment.

The crusade for peace was enthusiastically endorsed by the people of the Canadian West. Militarism and navalism were denounced by the *Guide* and the grain growers' organizations in numerous resolutions and memorials. The "follies and barbarism" of war were often pointed out to the prairie reader, and "native militarists," such as Sir Sam Hughes, were excoriated as "a positive menace to peace."[66] The costs of war, the inhumanity and the futility of armed conflict, were frequent themes of lectures, pamphlets, and articles by prominent grain growers. Antiwar propaganda in Canada emphasized similar themes and advanced the same arguments used in the United States before World War I.

Shielded as they were by three thousand miles of ocean, protected by British sea power and the Monroe Doctrine, and absorbed in the development of a great share of the continent, farmers on the Canadian prairies were often as isolationist in sentiment as their cousins in the American Middle West. Farmers' societies on both sides of the boundary passed similar antiwar resolutions during these years. The expenditure of public funds for military purposes was condemned as useless and wasteful. There was no excuse, thought many North American farmers, for squandering on arms and armies monies which were needed for the "country's development along natural lines."[67]

[65] *Grain Growers' Guide*, 4:5 (June 19, 1912). The "saw-off" was an agreement between the two regular parties in which one party consented not to run in a constituency so that a key member of the other party would have a seat in the House.

[66] *Ibid.*, 4:8, 9 (May 22, 1912).

[67] *Ibid.*, 7:5 (May 27, 1914).

Blinded by geographic isolation, many Canadian farmers were bitterly opposed to Canadian participation in the naval rearmament of the British empire. The *Guide* deplored the naval race between Britain and Germany and dismissed the "German menace" as a myth. War scares were believed to be the fabrication of the "vested interests," especially of the munitions manufacturers.[68] The prairie farmers thus anticipated the "merchants of death" theory which was later so widely believed throughout North America.

Western farmers were strongly opposed to the Canadian naval program which was adopted after months of bitter debate in 1910. "The common people," wrote the editor of the *Guide*, "have never asked for a navy," and the grain growers' associations supported his assertion with antinaval resolutions.[69] Since both political parties were supporting the naval armament program, the *Guide* believed that a referendum should be held allowing the people to decide the issue. This opposition placed the agrarians in alliance with the traditionally isolationist French Canadians. Both groups were denounced as traitors to the empire, but the French Canadians stood firm and the farmers remained convinced that the ideals of militarism were "utterly opposed to democracy."[70]

The crusade for peace enlisted the devotion of the ablest prairie radicals. They fought the drift toward war with every possible argument. Every evidence of militarism in Canadian life was exposed to ridicule and denunciation, and on one occasion the *Guide* even condemned the Boy Scouts as a militaristic organization.[71] As late as May 1914 the "Friends of Peace" protested against any form of military training in Canada and prophesied, ironically, "this militiamania epidemic will soon pass."[72] Only a few months before the fateful days of August 1914, and at a time when European statesmen were seeking desperately to avoid a world conflict, the *Guide* categorically declared, "Canada has never needed any Naval Protection, does not need any now, and has not asked for any."[73]

[68] *Ibid.*, 5:12 (September 4, 1912); "Who Organizes War Scares?" *ibid.*, 5:5 (September 18, 1912).

[69] *Ibid.*, 5:5 (December 18, 1912); *ibid.*, 6:5 (January 15, 1913); *ibid.*, 6:6 (January 29, 1913).

[70] *Ibid.*, 7:5 (May 27, 1914).

[71] *Ibid.*, 6:5 (September 10, 1913).

[72] *Ibid.*, 7:5 (May 27, 1914).

[73] *Ibid.*, 6:5 (July 23, 1913).

North American isolationism thus left many Canadian farmers as ill-prepared for war in Europe as were American farmers. In the four costly years that followed, however, Canadians learned lessons which many Americans were to learn a generation later in an even costlier world conflict.

The quest for the better life was interrupted when Canada became engulfed in the Great War in 1914. The farmers of the West, with the aid of their trade unionist allies, had challenged the plutocracy and had questioned many of the doctrines of the New Feudalism. They had marshaled and drilled their forces for the battles which were certain to come when peace in Europe again freed them to carry on their struggle for democracy in Canada. Their fear of the trend toward concentration in the economic life of the dominion and their deep faith in democracy had forced them to challenge the great and the powerful. If they lost the battle over the tariff, they won other, and perhaps more lasting, victories. If their achievements seem meager a generation later, it should be remembered that their work set the stage for the postwar revolt which in 1921 swept prairie radicalism into the very citadel of the New Feudalism in Ottawa. If their reforms seem born of political innocence, their wiser sons cannot claim to have created a better world than they planned.

V

The Nonpartisan League Invades Canada

The movement of the Nonpartisan League into western Canada is one of the most striking parallel developments in the history of American and Canadian farmers' movements. So close is the parallel that Sidney Godwin, editor of the *Non-Partisan Leader* and one of the foremost Saskatchewan leaguers, admitted that "the program and pledge is practically a duplication of the Nonpartisan League in North Dakota."[1] Such duplication is not surprising, for the league in Canada was born of the same grievances and advocated the same solutions as the parent organization in the United States.

Scarcely a year after A. C. Townley had begun his efforts to organize the Nonpartisan League in North Dakota, it appeared in Saskatchewan and quickly spread into neighboring Alberta. The league invaded the dominion when S. E. Haight returned from North Dakota in early July 1916 to his farm near Swift Current. Haight had worked with the league while in the United States, and with unusual zeal and persistence he undertook to spread the doctrine of the Nonpartisans and to organize a league among his Canadian neighbors.[2] Within a few weeks he had induced more than a hundred farmers to join the society, and as early as July 12 a meeting was held in Swift Current which elected Haight president and provided for a board of directors for the infant organization.

[1] Sidney Godwin, "The Farmers' Political League," *Grain Growers' Guide*, 9:19 (October 4, 1916).

[2] *Ibid.*; *ibid.*, 10:8 (September 5, 1917); Louis Aubrey Wood, *A History of Farmers' Movements in Canada*, 294. Mr. Wood is mistaken when he says, "S. E. Haight, a prominent member of the Nonpartisan League, having arrived in Saskatchewan from North Dakota, took up his abode at Swift Current." Haight had arrived from the United States some time before and was established near Swift Current when he journeyed to North Dakota for a visit during which he became interested in the league.

The league swept across the southern Saskatchewan plains like a prairie fire. Within two months the membership had reached twelve hundred; by February 1917 it had passed the two thousand mark; and by the following September nearly five thousand farmers were in the league.[3]

The growth in Alberta was almost as spectacular. Early in December 1916 five farmers familiar with the program of the league in Saskatchewan met in Calgary and organized an Alberta branch. With funds from dues contributed by the five charter members, the league financed its expansion and within two months two hundred farmers were enrolled as paid members.[4] Growth was so rapid that on February 14, 1917, the league held its first convention in Calgary. M. E. Sly of Strathmore was elected president and William Irvine began his remarkable political career as the league's first secretary. Under the leadership of President Sly the membership grew to nearly three thousand within a year.[5]

Though the league experienced a mushroom-like growth in these two provinces, it made slight headway in Manitoba. A few locals were organized in the Swan River Valley through the efforts of William Irvine, the enthusiastic secretary of the Alberta League,[6] but Manitoba farmers were extremely reluctant to forsake their strong Grain Growers' Association for so radical a group as the league. Moreover, traditional parties were more firmly entrenched in this older province. This situation created an atmosphere hostile to political innovation and discouraged the growth of political agrarianism. Later, when a farmers' party appeared in Manitoba, it retained so much of the psychology and appearance of the older parties that it was often charged with being an annex to the Liberal party.

The causes for the phenomenal growth in Alberta and Saskatchewan are strikingly similar to those which explain the sweep of the league south of the international line. Geographical propinquity and the ease with which men and ideas crossed state and international boundaries account for the rapid expansion into Mon-

[3] A League in Canada," *Nonpartisan Leader* (Fargo, N.D.), 4:9 (February 8, 1917); "The Nonpartisan League," *Grain Growers' Guide*, 10:8, 24 (September 5, 1917).

[4] *Alberta Non-Partisan* (Calgary), 2:10 (March 29, 1918).

[5] *Ibid.*, 5.

[6] *Ibid.*, 3:2 (April 23, 1919).

tana, Minnesota, South Dakota, and the prairie provinces. After crossing these political boundaries, the league program struck roots into the same economic and political discontent that had caused its birth in North Dakota.[7]

Basic economic grievances, rather than immediate financial distress, explain the friendly reception given league propaganda by farmers in the hard spring wheat belt. Though memories of the sharp depression of 1913 lingered, in 1916 farmers in both countries were enjoying a prosperity born of good crops and wartime prices.[8] The crop of 1915 was the largest yet produced in western Canada, and wartime prices guaranteed adequate returns. In spite of this temporary prosperity, the league's demands for economic reform appealed strongly to farmers in both regions.

The Nonpartisan League was more than a protest against high transportation costs, low farm prices, and unfavorable marketing conditions. It was a protest against underlying economic developments which many western farmers believed were evils inherent in a maturing capitalistic system.[9] The anti-monopolistic demands for the nationalization of public utilities, of banking and credit systems, of all industries "in which competition has virtually ceased to exist," and the extension of the public domain to include all coal mines, water power, and forests reflected the Socialist background of the league in North Dakota.[10]

It is obvious that the reforms advocated by the Nonpartisans went beyond those urged by earlier agrarian movements in Canada. This led to an invitation to Canadian Socialists to make common cause with the league to effect fundamental changes in the economic system. "The first logical step for the Socialist to take," wrote the editor of the official organ of the Alberta Nonpartisans, "is to ally himself with the Non-Partisan movement." [11]

The revolt against partyism in western Canada is a further explanation for the league's success. Traditional party spirit burned

[7] Roderick McKenzie, "How Farmers Captured a State," *Grain Growers' Guide*, 9:18 (September 6, 1916).

[8] Robert H. Bahmer, "The Economic and Political Background of the Nonpartisan League," 5, unpublished Ph.D. thesis, University of Minnesota, 1941.

[9] Theodore Saloutos, "The Rise of the Nonpartisan League in North Dakota, 1915–1917," *Agricultural History*, 20:43 (January 1946).

[10] "The Nonpartisan League," *Grain Growers' Guide*, 10:8 (September 5, 1917); Bahmer, *op. cit.*, 433–38.

[11] *Alberta Non-Partisan*, 2:7 (March 29, 1918).

low in both of these youthful provinces established only in 1905. The frequent artificiality of established parties was not obscured by long tradition as in the older provinces, and many prairie farmers were deeply concerned about political habits which in eastern Canada were taken for granted. By 1917 many who condemned the party system were seeking political expression in a "natural" or "nonpartisan" movement. The league was quick to point out that it was such an organization. Nonpartisans maintained that their program was "the crystallization into political action of an ideal of citizenship which has long been in the process of formation," and deserved the support of all who believed that partyism was doomed.[12]

League officials were fully aware of the strong appeal of the nonpartisan position, and the evils of partyism were emphasized to win support for the league where other appeals might have failed. "The party system with its hollow sham and hypocrisy, special privileges in legislation, a rotten civil service, narrow minded nationalism, and pretended patriotism of grafting politicians has got to go" declared the editor of the *Nutcracker*, and thousands of western farmers said amen.[13]

The prairie provinces were ripe for an agrarian political movement. The rank and file of the grain growers were restless after a decade of indoctrination against the traditional parties. For many years the *Grain Growers' Guide* had fed this political discontent by its constant emphasis upon the need for farmer representation in government and the urgency of political reform. On top of this, the consistent refusal of the organized farmers to enter politics created an unusual opportunity for the league. Nonpartisan leaders took full advantage of the rising demand for political action. They made every effort to present the league as the political phase of the already established farmer societies. League propaganda presented the United Farmers of Alberta and the Saskatchewan Grain Growers' Association as the educational aspects of the movement, the Grain Growers' Grain Company as the cooperative phase of the movement, and the Nonpartisan League as the political wing of the farmers' trinity of organizations.

[12] *Ibid.*, 1:8 (July 20, 1917).
[13] *Nutcracker*, 1:5 (July 6, 1917).

This propaganda made serious inroads into the established organizations, particularly the U.F.A. Locals of the U.F.A. and the S.G.G.A. endorsed the league as the political expression of the farmers' movement, and there is no doubt that the league received its chief support in both provinces from members of the older societies.[14] Just as the league in North Dakota and Montana capitalized on the demand for political action created by the Society of Equity,[15] so the league in Canada reaped a harvest of votes where others had sown.

The league undoubtedly made a strong appeal to thousands of ex-Americans in Alberta and Saskatchewan. The movement of Americans into western Canada had come principally from the border states which had a long history of agrarian agitation and political radicalism. Farmers from these areas were inclined to look with approval upon the program of the Nonpartisans. John W. Leedy, a former Populist governor of Kansas, Bert Huffman, an agrarian organizer and author whose career began with Coxey's famous army, S. E. Haight, an active leaguer in North Dakota, Daniel F. Boissevain, a former editorial writer for *Ranch* and an agrarian radical in Washington, and George Bevington, the money and credit expert, were prominent among the ex-Americans whose leadership did much to promote the growth of the league in Canada. Harry Johnson was another. His untimely death during the flu epidemic of 1918 unquestionably deprived the farmers' movement of one of its ablest leaders. The *Alberta Non-Partisan* believed that he was "probably most responsible for the success of the League in Alberta." [16]

The influence of the league was further strengthened when it became the vehicle for western Canadian radicals. Men interested in all phases of reform were drawn to the league, and they furnished it with a leadership that was unusual. Men whose names were soon to be synonymous with Canadian liberalism wrote their first articles for league journals and first raised their voices in defense of league doctrines. The most prominent of this group was the Alber-

[14] *Alberta Non-Partisan,* 3:7 (April 23, 1919); *ibid.,* 3:11 (May 22, 1919); *ibid.,* 3:7 (February 12, 1919); *Sun* (Swift Current), November 16, 1917.

[15] Robert H. Bahmer, "The Society of Equity," *Agricultural History,* 14:62 (January 1940); Theodore Saloutos, "The Montana Society of Equity," *Pacific Historical Review,* 14:408 (December 1945).

[16] *Alberta Non-Partisan,* 3:5 (November 6, 1918).

ta society's secretary, William Irvine, whose political career has spanned three decades and as many parties. Many Canadians, then and since, have considered him the A. C. Townley of the Canadian league.

The two liberal former professors of Wesley College in Winnipeg, the Reverend S. G. Bland and the Reverend J. S. Woodsworth, were deeply concerned with the progress of the league. Dr. Bland frequently contributed sermons and articles to league journals, and Woodsworth served as secretary of the Alberta league for a short period in 1918.[17] The golden-voiced orator, F. J. Dixon, an independent labor member of the Manitoba legislature, often supplied the oratorical punch to league-sponsored assemblies. His lectures, "Power of Ideals" and "How to get Rich without Working," were famous throughout the prairies.[18]

These men, and many others, gave the organization a wealth of able leadership that goes far to explain the influence of the league in western Canada. The People's Forum in Calgary presented many of the finest public speakers in Canada and served as a sounding board for the Nonpartisans.[19]

Political successes of the parent organization also served to publicize the league in Canada. The overwhelming victory in the North Dakota election of 1916 was electrifying news to western Canadian farmers. Roderick McKenzie, secretary of the Manitoba Grain Growers' Association, was immediately sent to study the North Dakota experiment. His published report was highly favorable and did much to advertise the program of the league.[20] The *Guide* commented approvingly and suggested that "so long as farmers of Canada or any other country are willing to vote blindly for either one of the old political parties, just so long will they remain a nonentity in shaping the legislation of the country."[21] Frequent articles in league publications in Canada from C. W. McDonnell, a member of the North Dakota legislature, kept Canadians informed of the progress of the Dakota society.[22] This pub-

[17] *Ibid.*, 2:5 (July 12, 1918). [18] *Ibid.*, 2:10 (April 2, 1918).

[19] *Nutcracker*, 1:13 (August 31, 1917).

[20] *Nonpartisan Leader* (Fargo, N.D.), 3:11 (September 14, 1916); *Grain Growers' Guide*, 9:18 (September 6, 1916).

[21] *Ibid.*, 9:5 (September 6, 1916); *ibid.*, 9:6 (November 29, 1916).

[22] For examples of these reports see the *Alberta Non-Partisan*, 3:3 (January 2, 1919); *ibid.*, 2:10 (March 24, 1918).

licity not only created interest in the league, but was an important factor in arousing Canadian farmers to the potential power of group action in politics.

Similar methods of organization, propaganda, and finance characterized the movement on both sides of the international boundary. Professional organizers were commissioned in both Alberta and Saskatchewan to spread the Nonpartisan gospel and extract the fifteen dollar fee as painlessly as possible. These men were invariably fluent talkers who could frequently "high pressure" a farmer into joining the league. The organizers received 25 percent of the fee if they furnished their own automobiles. If they drove one of the thirteen cars owned by the Saskatchewan league, or one of the eleven operated by the Alberta society, they received 20 percent of the fifteen dollars.[23] With this remuneration as a stimulus, league organizers combed the countryside, but always in the company of a well-known resident of the community whose presence served at once to introduce the organizer and to quiet any fears the farmer might have that he was being "played for a sucker."[24]

There was considerable criticism of so high a fee, but league officials defended the charge as necessary because of high organizational costs. "We had been warned by the parent organization," wrote Sidney Godwin, "that this would be necessary to cover the Saskatchewan mileage in canvassing, the per capita mileage of Saskatchewan being double that of North Dakota."[25] Alberta leaguers answered criticisms of the fifteen dollar fee with similar explanations, and the *Nutcracker* carefully pointed out that "the cost of organization in cities is practically nil, while in sparsely settled rural districts, the cost is very heavy."[26]

The practice of accepting postdated checks and the use of Ford cars were characteristic of league organizers everywhere. Unfortunately, the league lost heavily on both. Automobiles frequently had to be sold at a loss when organizing activities slackened, and many farmers either were unable to honor their checks or refused

[23] *Ibid.*, 2:5 (March 29, 1918); *Grain Growers' Guide*, 10:8 (September 5, 1917).

[24] Testimony of Mike McLachlan, prominent Nonpartisan of Swift Current, Saskatchewan.

[25] *Grain Growers' Guide*, 9:19 (October 4, 1916).

[26] *Nutcracker*, 1:5 (July 6, 1917).

to do so after the initial enthusiasm had worn off. The Canadian league realized that efforts to collect these bad checks through legal action would furnish critics with unfavorable publicity which could be turned against the society; therefore no effort was made to force payment.

League papers and journals were much the same in both countries. They were argumentative and highly partisan in their treatment of current issues. All employed a simplicity and directness of attack that was highly effective. Cartoons were employed extensively and Congressman John Baer's best cartoons were reprinted in Canadian papers without substantial change, the American Manufacturing Association becoming the Canadian Manufacturing Association and the Great Northern becoming the Canadian Pacific Railway in texts accompanying the pictures.[27]

The same catch phrases attracted the reader's attention in all league publications. "Big Biz" for industry, "Kept Press" for opposition journals, "We'll Stick" indicating the farmers' resolve to stand together, "Peanut Politicians," "Politics is the business of the people," and "Go home and slop the hogs" were the catch phrases most frequently used. Of these "Go home and slop the hogs" was unquestionably the most effective. Supposed to have originated with Treadwell Twichell in North Dakota, it was often quoted as the actual utterance of this or that local politician, thus reflecting the keen resentment many farmers felt toward their legislators. It was used so often that it soon became a colorful part of the oral tradition of North American agrarianism.

The league in Canada produced a well-edited and unusually effective press of its own, primarily because of the conviction that only in Nonpartisan papers could the truth be told. The "kept press" of a capitalistic society was frequently dishonest and always prejudiced. Only the official papers, the *Non-Partisan Leader*[28] and the *Alberta Non-Partisan*,[29] could be relied on to present the "nonpartisan" truth. Both papers had an influence outside the league, for in both provinces the circulation exceeded the league's

[27] For examples see the *Alberta Non-Partisan*, 1:1, 9 (October 12, 1917); *ibid.*, 1:9 (October 25, 1917); *ibid.*, 2:9 (January 4, 1918); *ibid.*, 3:9 (March 26, 1919).

[28] Published in Swift Current, Saskatchewan (circulation 5000).

[29] Published in Calgary, Alberta (circulation 9500). See the *Alberta Non-Partisan*, 2:13 (June 21, 1918).

membership. A few Canadian leaguers subscribed to the *Nonpartisan Leader* of Fargo, North Dakota, along with their provincial paper, thus keeping informed of developments south of the forty-ninth parallel.[30]

To prevent opposition in the early stages of organization, the league avoided publicity concerning its activities.[31] A. C. Townley's organizers had ostensibly been "selling washing machines," and Canadian organizers used similar cover. Vague rumors, of course, circulated throughout the region that something was afoot, but the nature of the society remained a secret until the league felt strong enough to sustain the attacks certain to be hurled against it. One Saskatchewan editor warned his readers that "a band of political gypsies" had invaded their province, but that they could be recognized easily for "they have $15 marks on their eyelashes, and their focused vision is an automobile."[32]

The reforms advocated by the Nonpartisan League covered a wide range of grievances and touched nearly every important phase of Canadian life. Because of the constitutional limitations placed upon provincial powers by the British North America Act, the league program was chiefly aimed at federal action. Where provincial legislation could be used effectively, however, the league supported it.

The proposed economic changes were fundamentally a demand for the "social service state" and foreshadowed much that later appeared in the platforms of the Social Credit and Cooperative Commonwealth Federation parties. Frequent resolutions of the league conventions made it clear that sweeping economic reforms were contemplated through the nationalization of public utilities, the elimination of monopolies through government ownership, and the institution of graduated income and inheritance taxes. The debtor West would get cheap money through the nationalization of the banking and credit systems, and the removal of the protective tariff would solve the farmers' remaining economic problems. The elimination of the middleman and the land speculator would be necessary to satisfy the principle that "what the producer produces

[30] *Nonpartisan Leader* (Fargo, N.D.), 4:9 (February 8, 1917).

[31] Godwin, "The Farmers' Political League," *Grain Growers' Guide*, 9:9 (October 4, 1916).

[32] *Sun* (Swift Current), August 18, 1916.

shall go to the consumer direct through government channels and the manipulator of prices — the easy money guy — be thus eliminated."[33]

Though economic proposals were the heart of the league platform, political reform was also demanded. Nonpartisans shared the implicit faith of the early twentieth century that the evils in democracy could be eliminated through the use of direct legislation, the initiative, referendum, and recall. All Nonpartisan candidates for political office signed undated recall statements to be used in the event they violated league principles while in office.[34] Thus Nonpartisans were the first to adopt the recall in Canadian politics. Electoral changes — including the publication of election expenses, the designation of election days as public holidays, the announcement of provincial elections ninety days before the voting, the abolition of the election deposit, and a better system of compiling voters' lists — were advocated to purify politics and remove the political handicaps so keenly resented by western Canadian farmers.

Political patronage and graft got particularly bitter criticism from the league. "The patronage system," declared William Irvine, "is one of Canada's social evils — a canker, a disease more blighting, more demoralizing than any other social disease that infects the body politic today." Equally vicious was the party slush fund, "which has perverted the public mind, made representative government in Canada a machine, and enabled the Canadian politicians to exploit the nation during the last century and more."[35]

The only guarantee against these evils, in the opinion of the league, was an efficient "business government" responsive to the will of the people.[36] The Alberta league explained that it did not wait for "a platform to be nailed together in Ottawa or Edmonton by a few deeply interested politicians. Its program is drawn up and endorsed by the assembled electorate. A man is selected to

[33] *Alberta Non-Partisan,* 2:9 (November 23, 1917); *ibid.,* 2:8 (March 15, 1918); *ibid.,* 2:7 (January 4, 1918); *Nutcracker,* 1:12 (February 3, 1917); *Grain Growers' Guide,* 10:24 (September 5, 1917); *Alberta Non-Partisan,* 2:8 (January 4, 1918); *ibid.,* 2:5 (June 7, 1918).

[34] *Ibid.,* 2:9 (November 23, 1917); *ibid.,* 1:8 (November 9, 1917); *ibid.,* 2:6 (February 8, 1918).

[35] *Nutcracker,* 1:6 (February 3, 1917).

[36] *Alberta Non-Partisan,* 3:8 (May 22, 1919).

secure the legislation desired, the expenses incurred in elections are paid directly from the pockets of the electors, a balance sheet is submitted each year, and the books of the League are open for investigation by any member."[37]

A few leaguers apparently believed that "business government" could be accomplished by a commission form of government similar to that adopted by several American cities during this period.[38] This idea had been discussed in Canada for several years and was in line with Sir John A. Macdonald's concept of the position of the provinces in the unitary form of government he envisioned.

The league took a strong stand on the question of equal rights for women and advocated their social and political equality at every opportunity. "The Nonpartisan platform is the only place where the high minded women of this Province can stand with dignity and clean feet," asserted the editor of the *Nutcracker*, and when Mrs. Louise C. McKinney entered politics on the league ticket, Alberta Nonpartisans proved the sincerity of their position by electing her to the legislative assembly.[39] The Nonpartisan League thus won for itself the distinction of electing the first woman to a legislative seat in the British empire.

Social reforms also attracted the attention of the league. Resolutions calling for the abolition of the liquor traffic were frequently passed and Mrs. McKinney served as president of the Alberta W.C.T.U. for several years. Opposition to capital punishment was also voiced.[40] National compulsory insurance covering accident, illness, and old age was urged to relieve the distress caused by these ancient enemies of mankind. The "Spanish Flu" epidemic in 1918 furnished the league, on both sides of the international boundary, with an opportunity to point out the need for "nationalized medicine."[41]

Certain changes advocated by the league, of course, were entirely peculiar to Canadian institutions and had no parallel in the program of the American league. The Nonpartisans in Canada

[37] *Ibid.*, 1:8 (September 29, 1917).

[38] *Nutcracker*, 1:5 (June 7, 1917).

[39] *Ibid.*, 1:5 (June 21, 1917).

[40] *Ibid.*, 1:15 (February 16, 1917); *Alberta Non-Partisan*, 2:7 (January 4, 1918); *Grain Growers' Guide*, 10:8 (September 5, 1917).

[41] *Alberta Non-Partisan*, 2:9 (November 23, 1917); *ibid.*, 2:5 (November 20, 1918).

found themselves seeking power in a political system based on the two-party principle and the power of the cabinet to accept any issue as a test of strength.[42] The league, quite naturally, was bitterly opposed to "cabinet domination" as irreconcilable with its own conceptions of popular sovereignty. "In a business administration as advocated by the league, the proportional idea would be carried into the cabinet; the life of such a government would not be dependent on the defeat of any measure whatsoever; the public business would be thoroughly departmentalized and directed by experts, while every measure would be passed or defeated by the assembly on its merits."[43] It was clearly understood that if a ministry could demand a vote of confidence on any issue, government by group representation, as advocated by the league, was an impossibility.

Another Canadian institution which drew criticism from the league was the practice of awarding titles to distinguished citizens. "Canada . . . is rapidly falling to the miserable level of sycophancy and greed for titular appelation. The absence of title among our public men will soon carry a far greater distinctive honor than the titles" was the opinion of one Nonpartisan, and the official pronouncements of the league reveal that his opinion was that of the majority.[44]

The league took a most radical stand on the vexing question of the reform of the Canadian Senate. Most liberal and agrarian groups advocated reform, but the league believed that the only solution was the abolition of the Senate as an archaic and undemocratic institution. Nonpartisans also maintained that no court should be legally competent to declare any act of Parliament unconstitutional.[45]

In view of the comprehensive reforms advocated by the league, it is not surprising that many Canadians were shocked, and that charges of Bolsheviki, radical, and Red were hurled at Nonpartisans.

The Canadian league felt a strong sympathy for labor and

[42] W. L. Morton, "Western Progressive Movement and Cabinet Domination," *Canadian Journal of Economics and Political Science*, 12:137–38 (May 1946).

[43] *Alberta Non-Partisan*, 3:8 (May 22, 1919).

[44] *Nutcracker*, 1:13 (March 2, 1917); *Alberta Non-Partisan*, 2:9 (November 23, 1917); *ibid.*, 2:14 (June 7, 1918).

[45] *Ibid.*, 2:9 (November 23, 1917).

maintained close relationships with its leaders. This was in line with the nature of the economic reforms advocated by the league and was encouraged by the universal interest in reform shared by league officials. The struggle against "entrenched wealth" was too great a task for the farmer alone. The cooperation of every producers' group was essential. Editorials in league journals supported labor's position, and independent labor candidates were given publicity along with league candidates.[46] When the Labor Representation League nominated William Irvine as its candidate for South Calgary in the provincial elections of 1917, the Nonpartisan League gave enthusiastic support.

Objections were raised against this alliance by certain Nonpartisans who believed the league should be a farmers' party. In a sense this conflict foreshadowed the famous "broadening out" issue which split the ranks of the organized farmers in later years. The league avoided the exclusiveness which characterized later farmers' political movements. League officials were reluctant to exclude members on the basis of employment or occupation. They feared that limiting the organization to one class would eventually become a source of weakness rather than strength.[47] The worker and the farmer were allies. Failure to cooperate would mean that victory would go to "the interests."

The league was less friendly toward the established farmers' organizations. Criticism of existing societies was inevitable in the competition for the farmers' support. Rivalries were kept in the background at first, in the hope that the U.F.A. and the S.G.G.A. would adopt the league as the political wing of the organized farmers.[48] When the league was eventually forced to abandon these hopes, many Nonpartisans bitterly turned against the organized farmers.

The failure of the organized farmers to enter politics was the focal point of the attack. The *Alberta Non-Partisan* admitted that the United Farmers' movement was a blessing, but "because their organization calls for non-interference in the political field . . . this vast body of producers is at the mercy of every little punk politician on the lot." The S.G.G.A. was condemned as nothing

[46] *Ibid.*, 2:8 (July 12, 1918).
[47] *Ibid.*
[48] See above, p. 80.

more or less than a Liberal organization, ruled and dominated by the Liberal party. The "Farmers' Platform," framed by the grain growers, was looked upon as a noble but futile gesture. Though "thoughtfully made up, broad and democratic in outlook," it was meaningless, for it was "a platform without any person to stand on it." The *Grain Growers' Guide* was condemned as being reactionary, and "as plutocratic an organ as the [*Toronto*] *Globe*."[49]

Savage attacks were launched against the officials of the older organizations. One league writer bluntly charged that "with few exceptions, neither its leaders nor the men who compose its membership had any motive higher than the egoistic one of improving their own position, or moulding political conditions so as to satisfy their own desires."[50] Clearly, there were those who believed the league would gain by discrediting the grain growers' movements. This method won a few converts, but it created a suspicion and ill will which the league never outlived. In Saskatchewan it gave rise to a bitter resentment which prevented the two organizations from ever coming to terms.

Criticisms of the grain growers were somewhat moderated by the influence of leaguers whose primary loyalty remained with the older societies. Prominent leaguers such as James Weir, D. J. Sykes, Rice Sheppard, and J. C. Buckley made every effort to heal the divisions which inevitably appeared. They sought to reconcile differences by emphasizing the educational functions of the one and the political purposes of the other. "Whether owned or disowned, the Non-Partisan League is the child of the U.F.A.," declared William Irvine in a Dominion Day address before hundreds of U.F.A. members. Subsequent events proved that his charge could not be denied.

[49] *Alberta Non-Partisan*, 2:8 (December 7, 1917); resolution of the Nonpartisan League, February 23, 1917, *Sun*, February 27, 1917; *Alberta Non-Partisan*, 3:5 (January 2, 1919); *ibid.*, 1:6 (October 25, 1917).

[50] C. L. Carp, "Fate of the Grain Growers," *Alberta Non-Partisan*, September 27, 1918.

VI

The Nonpartisan League in Prairie Politics

The provincial elections of 1917 revealed the extent of the sweep of the Nonpartisan League through Saskatchewan and Alberta. The league was still in the early stages of organization when the opportunity came to prove its strength in both provinces. It was compelled to enter political campaigns under its own banner and as a distinct party.

The parliamentary system of responsible government could not be reconciled with the nonpartisan concept, nor could the leaguers capture the older party organizations as they had in North Dakota, for there were no direct primary elections. Candidates were selected by party conventions, and parliamentary procedure discouraged an independent legislative attitude. Wherever the league entered candidates in Alberta and Saskatchewan, it was forced to campaign as a third party. The Canadian league thus lost the initial advantage its counterpart to the south possessed.

The test of strength in Saskatchewan came in the provincial election of June 1917. The league entered the campaign with confidence and high hopes; it emerged from the election broken and dispirited. Eight candidates were nominated by enthusiastic but poorly disciplined conventions. All the candidates were farmers, and the two who ran the strongest races were prominent members of the grain growers' movement.

Serious mistakes were made throughout the entire campaign. The nominating conventions were unruly and badly managed. An ill-conceived policy of secrecy quickly antagonized the public press, and the high-handed methods used by certain officials alienated many friends. League conventions were frequently torn by strife and personal feuds. Convention programs were poorly organized, and many delegates were offended by the attitude of some speakers toward the S.G.G.A. Ineptitude and incompetence foreshadowed the debacle which followed.

Victory was achieved in only one of the eight ridings, and it was a hollow one for the league. David J. Sykes of Swift Current, a self-styled "independent Liberal since 1911," was so popular that he won the nomination of all three parties and was elected by acclamation.[1] Leaguers were cheered by the thought that they had nominated Sykes before the Liberals and Conservatives had acted, but this added little to the victory. Among the other league candidates only Mrs. S. V. Haight in Thunder Creek made a strong showing. Nicol McVean in Happyland, J. Cameron in Last Mountain, E. W. Grainger in Morse, and W. S. Simpson in Notukeu ran such poor races that they forfeited their election deposits.

This disappointing outcome shattered the league's morale in Saskatchewan. There followed a period of recrimination and acrimonious debate which weakened the league and blighted its future. Charges of disloyalty and bribery were followed by wholesale expulsions from the organization.[2] The growing discontent came to a head in the annual convention in Swift Current on July 11, 1917.

The stormy sessions of this convention marked the decline of the Saskatchewan league. Its subsequent strength was more apparent than real. Angry charges of "liar" and "traitor" were often heard as delegates fought over the dying organization. P. L. Craigen, W. A. Gambie, and Percy Hooper were formally tried for aiding political enemies of the league during the recent election. Craigen, who previously had been "expelled" by the league executive, led the fight to reform the league.[3] Efforts to rid the organization of "bureaucracy" and "executive dictatorship" had this measure of success: The convention elected a new executive and appointed a committee to draft a new constitution. Craigen, however, was disgusted and advised farmers "to keep their $15 in their pockets" until the league had an honest, efficient financial administration.[4]

[1] *Sun* (Swift Current), March 30, 1917.
[2] *Ibid.*, July 13, 1917; *Grain Growers' Guide*, 10:19 (March 21, 1917); *ibid.*, 10:16 (October 10, 1917).
[3] *Sun*, July 13, 1917.
[4] Letter of P. L. Craigen, *Grain Growers' Guide*, 10:16 (October 10, 1917).

Reform came too late to save the influence and prestige of the league. Sykes resigned from the society and thereafter acted as a Liberal in the Saskatchewan Assembly. The league, however, refused to accept his resignation, demanding that he honor the recall statement he had signed before the election. Sykes sidestepped this touchy issue by insisting that the recall pledge was not effective unless a petition signed by more than half the electorate could be presented.[5] Thus the first efforts to use the recall against a Canadian legislator failed.

League prestige was irreparably damaged by the disgraceful performance which followed defeat. The poor showing was clearly due to faulty planning, inept leadership, the disloyalty of prominent Nonpartisans, the adverse town vote, and the superior organization of the older parties. The purge which followed the election failed completely to revitalize the organization. Within a month the society was in desperate financial straits with less than four hundred dollars available and fifteen thousand dollars in post-dated checks outstanding.

Further league activities in Saskatchewan were anticlimactic. In the excitement of the federal election in December 1917 the Saskatchewan league nominated three provisional candidates, but it was too weak to enter any one of them in the campaign.[6] Many Nonpartisans later believed that the "Win-the-war" campaign killed their organization.[7] This was only partially true. The league was already moribund when the election was called.

The league was more successful in Alberta where the disastrous mistakes of the Saskatchewan society were avoided by a more discerning leadership. The Alberta league put less emphasis on "political principles" and more on leadership. With men like J. H. Ford, M. E. Sly, H. W. Johnson, James Weir, J. W. Leedy, and William Irvine in responsible positions, the Alberta league was much stronger than the Saskatchewan society.

Four months after its first convention the league in Alberta was faced with a provincial election. Four candidates — James Weir, J. W. Leedy, J. F. Miller, and Mrs. Louise C. McKinney — were

[5] *Sun*, July 13, 1917.
[6] The three were D. J. Sykes, J. G. Laycock, and Fred Laidley.
[7] Personal interviews with ex-leaguers in Saskatchewan.

endorsed.[8] The election was bitterly fought. League strength was an unknown quantity, and the victory of James Weir in Nanton and Mrs. McKinney in Claresholm was a surprise.[9]

The league was jubilant over the results. With pardonable pride, the editor of the *Nutcracker* exulted, "with but a few months organization work the Non-Partisan League was able to put four candidates in the field and was successful in electing two of the four!"[10] Friends of the league were optimistic about the organization's future, and its enemies feared that the election portended a sorry future for the province.

The election taught the league many important lessons. It was obvious that the society had been forced into an election many months before its organization was perfected. League candidates had been the object of ridicule and abuse. Nonpartisans learned the bitter truth that rumors and whispered innuendo were frequently more influential than the avowed issues of a campaign. League officials were denounced as dictators, Reds, and Bolsheviks, and the organization was charged with being "undemocratic," "a racket to bamboozle the farmer," and a "foreign importation."[11] The *Non-Partisan* cautioned leaguers to "let all rumors originate in our open conventions" and vigorously defended the democratic nature of the league. "The men who are engaged to organize the Non-Partisan League are paid for bringing the movement before the people. They are not your bosses. They do not shape your policies."[12]

The league's newly elected legislators were carefully watched by Alberta voters. Nonpartisans were delighted with their performance in the legislative assembly, while more objective observers thought their work useful but hardly distinguished. Weir was the abler of the two, and he kept the Liberal government under fire with charges of alleged graft, favoritism, and "rotten politics." Mrs. McKinney's frequent addresses before U.F.A. and

[8] William Irvine ran as an independent labor candidate in South Calgary but was supported by the league. Weir at the time was second vice-president of the U.F.A.

[9] Miller polled within seventy-three votes of the victor, but Leedy, with only a few days to organize a campaign, made a poor showing.

[10] *Nutcracker*, 1:8 (July 20, 1917).

[11] Practically all the newspapers were opposed to the league and attacked it editorially. The *Calgary Herald, Edmonton Bulletin,* and others kept the league under constant criticism.

[12] *Alberta Non-Partisan*, 1:8 (October 12, 1917).

W.C.T.U. groups also gave wide publicity to league activities in government.

World War I had a profound influence on the history of the league, both in the United States and in Canada. The entrance of the republic into the conflict in April 1917 placed the American league in an embarrassing and dangerous position. The demand for economic and political reform could hardly be abandoned if the league was to survive, yet continued emphasis upon these proposals would certainly create tensions and disunity dangerous in wartime. The result was inevitable. Charges of disloyalty and un-Americanism discredited the league and hastened its eventual collapse.

The Canadian league had a similar experience. Its Socialistic platform, especially its advocacy of a policy of conscripting wealth and raw materials, aroused bitter opposition. Its attitude toward the war unquestionably created deep suspicions and hatred. Few could reconcile the league's open criticism of Allied war aims with a vigorous prosecution of the war. The conflict was viewed by many Nonpartisans as a struggle against tyranny and "Kaiserism," and enemies of democracy at home were often more dangerous than the Kaiser's army. The belief that "the war in Flanders has endangered rather than advanced democracy"[13] could hardly be expected to win the warm support of relatives and friends of soldiers overseas. Canadian war aims were severely criticized as "vindictive," and "the seeds of another war." "We want to know," wrote a prominent leaguer, "how we shall know when the war is won. Are Canadians to keep dying until the war council in London cries enough? . . . To 'Beat Germany to her knees' or 'Crush Germany' would not be worth the blood of one Canadian as an end."[14] Sentiments like these were considered defeatist by many Canadians and outright sedition by others.

Nonpartisans angrily condemned war profiteering and graft. The *Non-Partisan* was convinced that "to hang one grafter at home would be of as great advantage as victory over the enemy abroad."[15] Though opposed to the conscription of farm labor, the league strenuously advocated the conscription of wealth to fight

[13] *Ibid.*, 1:5 (September 29, 1917).
[14] *Nutcracker*, 1:8 (August 17, 1917).
[15] *Alberta Non-Partisan*, 2:5 (January 4, 1918).

the war. "The urgent need of the moment, is a courageous, comprehensive and thorough National Service policy . . . one that will fearlessly scourge the money changers from their seats in high places in a manner worthy of the sacrifices rendered on the altar of the great God of war."[16]

This issue became increasingly important as the league grew in size. Through resolutions, editorials, speeches, and letters to officials in Ottawa, Nonpartisans pressed their doctrine. Failure to conscript national wealth was clear evidence to many leaguers that the party system was too corrupt and inefficient to handle wartime problems and that the wealthy interests of Canada "controlled the legislature."[17] They predicted further wartime restrictions, but feared these would only be "the continuance of conscripting the wealth of the people for the benefit of Big Interests . . ."[18] The dominion program of fixed prices for wheat was viewed in this light. Frequent protests by the executive of the Alberta league pointed out the injustice of fixing the price of the farmer's wheat without controlling the price of industrial goods.[19]

League enemies in Canada cited the alleged pro-German record of "Boss" Townley and the North Dakota league to embarrass the Canadian organization.[20] These charges were refuted in two ways. Evidence from North Dakota was produced to prove the loyalty of the parent society, and any intimate contact with the American organization was denied. The force of the latter argument was frequently weakened by the presence in league papers of lengthy articles from the Dakota society, often placed on a page opposite the article denying relationships with the Townley group.[21]

The league denounced the Borden government for "party politics" and frequently exposed lurid "war scandals." The refusal of H. W. Wood, president of the U.F.A., to accept a proffered cabinet position was applauded. The *Non-Partisan* was confident that "as the most capable and outstanding leader in Alberta, he would have been used to cover the crimes of the Borden government."[22]

[16] *Nutcracker,* 1:9 (February 3, 1917).
[17] *Ibid.,* 1:12 (August 31, 1917).
[18] *Alberta Non-Partisan,* 2:5 (January 4, 1918).
[19] Letter to Sir Robert Borden, reprinted in *ibid.,* 1:7 (October 24, 1917).
[20] *Calgary Herald,* July 2, 1918.
[21] *Alberta Non-Partisan,* 2:5 (July 5, 1918).
[22] *Ibid.,* 1:5 (October 26, 1917).

The government's decision to form a union cabinet and go to the people for approval in a "Win the War" campaign was a severe blow to the Alberta league, and the movement toward a bipartisan government was met with hostility, though the league had long advocated a similar concept of no-party government. Thomas A. Crerar, the foremost farm leader in western Canada, was censured for joining the cabinet and accused of having "sold out for the mess of pottage."[23] League officials announced that if the union government entered candidates against Nonpartisans, it would be "positive proof that the Union government is not sincere in its professed desire to get rid of partyism."[24]

Four league candidates sought election in the campaign of December 1917. D. H. Galbraith, president of a U.F.A. cooperative, was selected in the Bow River constituency, and Stephen T. Marshall, also a prominent member of the U.F.A., represented the league in Macleod. The perennial candidates, J. W. Leedy and William Irvine, once again sought election, though this time in Victoria and East Calgary.

The campaigns in these four constituencies were as fiercely contested as any in the election. The league was attacked with every possible charge, and animosities were created which lasted many years. Charges of Bolshevism, pro-Germanism, and Kaiserism were common. League candidates were subjected to a campaign of gossip and rumor inspired by fear of the radical platform of the league. George P. Smith, the Liberal candidate in Victoria, produced letters from Leedy's political enemies in Kansas which purported to prove that Leedy was dishonest, disloyal, and untrustworthy.[25]

Though the league vote cut deeply into Liberal strength in many areas, in no constituency was a Nonpartisan successful.[26] The appeal to emotion, patriotism, and necessity made by union candidates was a powerful force which overcame all opposition. Nonpartisans attributed their defeat to unpreparedness, "the unfair

[23] *Ibid.*, 2:5 (July 12, 1919).
[24] *Ibid.*, 1:5 (November 9, 1917).
[25] Smith withdrew late in the campaign to become provincial minister of education.
[26] The *Sun* (June 8, 1917) observed, "Wherever these Non-Partisan candidates run, they cut into the Liberal vote, and the Conservatives generally vote with their party."

election act," and the expenditure of enormous funds by the old parties,[27] but defeat did not destroy the spirit of the organization, as it had in Saskatchewan. The *Non-Partisan* admitted that the "election interfered with our organization," but thought farmers had learned that "they must be determined to work all the harder for the future, to avoid cleavage, for undermining tactics will always be used by financial interests."[28] Despite the defeat optimistic predictions of a promising future were made, and believed, by leaguers in all parts of the province.

A movement was already under way which was to dash these high hopes. Leaguers who were also members of the U.F.A. had grave doubts about the wisdom of dividing the farmers' strength. A strong sentiment was growing that the two organizations must join forces. The rank and file of the U.F.A. were now anxious to test political action, and it was increasingly evident that a reluctant U.F.A. leadership must eventually take the plunge.

Nonpartisans were badly divided on the wisdom of unification. Many believed that "it should be possible for the farmer to be both at the same time."[29] A stronger faction, led by James Weir, realized that the existence of two farmers' organizations was wasteful duplication and sought amalgamation to avoid a disastrous struggle. Still others hoped that the older group could be convinced that entry into politics would be suicidal and that political action should be left to the league. There were grounds for this hope in the opposition to political action by H. W. Wood, S. S. Dunham, and other influential leaders of the U.F.A. Then there was the minority of leaguers who hoped to retain the identity of the league and avoid all association with the larger organization. These "bitter enders" fought a rear-guard action against amalgamation until the issue was finally resolved.

The union of the two organizations was precipitated by the action of the annual convention of the U.F.A. in January 1919. The growing demand for political action was recognized in this convention to the extent that locals were authorized to enter politics in each federal constituency. This was clearly a compromise meas-

[27] *Alberta Non-Partisan*, 2:3 (December 20, 1917).
[28] *Ibid.*, 2:8 (December 20, 1917).
[29] John Glambeck in *ibid.*, 3:7 (January 3, 1919).

ure designed to satisfy both factions in the U.F.A. As a concession to majority opinion it possessed many weaknesses, but it did force the league to a decision. The *Non-Partisan* clearly understood that a momentous decision faced the Alberta farmers. "The real question to be considered is shall the political action in the future by the farmers of Alberta be taken through the U.F.A. or the Non-Partisan League?"[30]

The U.F.A. was forced into politics to preserve its organization. U.F.A. locals during 1918 and 1919 endorsed the league in increasing numbers, and sympathy for the league threatened to disrupt the program of the U.F.A.[31] By thus forcing the U.F.A. to act, the league signed its own death warrant. Once the decision had been made, the older and more powerful organization inevitably became the political voice of Alberta farmers. The process of absorption was stormy and involved and nearly failed on several occasions.

Nonpartisan officials were at first hopeful that the life of the league could be preserved. Their hope lay in the nature of the political structure erected by the U.F.A. convention. If the conventions could be induced to endorse the league, all would be well. The league, therefore, charted a suggested course of action for the U.F.A. in the form of a proposed resolution: "Resolved, that we, the U.F.A. believe in organization for political action, and that whereas we are an industrial organization . . . we commend to our members that political movement known as the Non-Partisan League, which is the true child of the U.F.A."[32] This action, promised league officials, would save the U.F.A. from splitting into two factions and would "solve the pressing question for farmers of what to do about politics."[33] They quickly discovered that such self-denial was too much to expect from the wealthier, larger, and older organization.

The first significant step toward unification was taken on January 17, 1919, when league officials, in a letter to H. W. Wood, proposed that a committee from the U.F.A. meet with the league

[30] *Ibid.*, 3:5 (March 12, 1919).

[31] *Ibid.*, 2:5 (January 18, 1918); *ibid.*, 3:3 (April 12, 1919); *ibid.*, 3:9 (April 23, 1919); *Manitoba Free Press*, July 15, 1919.

[32] *Alberta Non-Partisan*, 3:5 (January 15, 1919).

[33] *Ibid.*

executive, "to intelligently direct the independent political action of this province." As a result of this proposal, the executives of the two organizations convened in March to study the problem of joint action. The results were encouraging, and the *Non-Partisan* was convinced that "there is every likelihood of both being merged into one great political movement." The journal urged caution, however, until the U.F.A. agreed to enter provincial as well as federal politics, and until "the mass of the U.F.A. go into politics wholeheartedly."[34]

The U.F.A. representatives deferred the issue of union on the grounds that the organization's constitution placed such decisions in the hands of the locals. They were willing, however, to join in endorsing a resolution which encouraged the "use of every influence to bring about joint and cooperative action between these two forces for the attainment of those democratic ideals for which both organizations stand." A committee was appointed to "seek as far as possible to harmonize and unify all political action on the part of the Non-Partisan League and the U.F.A."[35]

The issue was decided in local conventions of the U.F.A., where struggles for power often came close to the surface. The *Non-Partisan* believed that "each should be willing to lose itself if necessary, in the interest of the common good," but this was a counsel of perfection. Responsible officials of the league showed a commendable spirit of moderation, though there was resentment that "when the League had demonstrated the possibility of successful political action, the U.F.A. decided to do what in all the years previous it had consistently refused to do . . ."[36]

The league executive committee cleared the way for absorption into the local conventions when on May 22, 1919, it issued a formal statement to all league members: "We therefore advise that our members amalgamate with the U.F.A. and see that the work the League undertook to do will be accomplished more speedily and with greater effect than if we stay separate."[37] This difficult decision was reached despite the admittedly strong temptation to fight the U.F.A., and thus preserve the life of the league. The

[34] *Ibid.*, 3:8 (April 8, 1919); *ibid.*, 3:5 (April 8, 1919).
[35] *Ibid.*, 8, 9.
[36] *Ibid.*, 3:5 (March 12, 1919).
[37] *Ibid.*, 3:9 (May 22, 1919).

U.F.A. owed a great debt of gratitude to J. C. Buckley, J. H. Ford, and other Nonpartisan leaders who possessed the wisdom and the will to avoid that costly struggle.

The path to union was not a smooth one. U.F.A. conventions in Macleod, Bow River, and Red Deer constituencies were the scene of bitter struggles. Leaguers sought to preserve as much of their program as possible in the new political associations. H. W. Wood and his associates, on the other hand, fought the league on every issue. Wood refused to endorse league principles and openly denounced the league as destructive to "class mobilization" and "class efficiency."

The U.F.A. political associations born of these conventions united the six thousand members of the league and the nineteen thousand members of the U.F.A.[38] In Macleod the political society retained the word "League" in its title. In East and West Calgary, league influence, through the work of William Irvine, was especially strong. In Red Deer the league demand for "business government" was altered to a demand for "honest and efficient" government.[39] Wherever impasses threatened the progress of the conventions, it was usually James Weir who fashioned an acceptable compromise. It is doubtful that successful union could have been achieved without the work of this peacemaker.

The *Non-Partisan* generally applauded the actions of the conventions, although as it witnessed the league's gradual absorption there were regrets that the U.F.A. retained its class basis. "There is no fundamental democratic principle which the farmers or any other industrial unit have as a monopoly" declared the editor. He also warned that "if the people in the towns, villages and cities can only vote for farmers' candidates, without having something to say regarding the political policy, and the selection of a candidate, it is not likely the farmers' party will be any more in demand than Tory or Grit."[40]

Despite these misgivings, the *Non-Partisan* enthusiastically endorsed the absorption of the league by the U.F.A. political associations. "There is now no excuse for our farmers to be outside the

[38] League officials estimated, however, that 90 percent of the league's members were also members of the U.F.A.

[39] *Grain Growers' Guide,* 12:14 (June 25, 1919).

[40] *Alberta Non-Partisan,* 3:5 (June 19, 1919); *ibid.,* 3:6 (June 5, 1919).

U.F.A. movement" declared the journal. "The U.F.A. represents, both politically and commercially, the collective opinions of the organized farmers of Alberta."[41] Later editorials urged "100% support" from league members for the U.F.A. political program, and the final issue of the *Non-Partisan* predicted, "The U.F.A. can elect a farmers' candidate in every rural constituency in Alberta in spite of both parties."[42] Thus, with a prophecy and a challenge, the Alberta league ended its career to enter a larger and greater destiny.

The Saskatchewan league never came to terms with the S.G.G.A., as the Alberta league did with the U.F.A. During 1920 efforts were made to revive the organization. In some areas, league organizers were mistaken for S.G.G.A. agents and were paid fifteen dollars by unsuspecting farmers.[43] Officials of the S.G.G.A. locals were alarmed by this and appealed to the provincial secretary, J. B. Musselman, for advice. Musselman accurately judged that the league would never regain its strength, and wisely counseled that persecution would only serve to attract attention to the impotent organization.[44]

The *Grain Growers' Guide*, however, professed to see great danger in the organization. The editor, George F. Chipman, excoriated the league as "an institution lacking both in 'pride of ancestry and hope of posterity'" and warned that "judging from the past, the receipt for $15 is about all the farmers may expect from their investment."[45] These editorial blasts were wasted, for the farmers of Saskatchewan had forced the Liberal government to become a farmers' government in all but name, and the attraction for the league was forever gone. They had chosen cooperation rather than government ownership and the league now represented a radical position which the majority of farmers were not prepared to accept.[46]

The Nonpartisan League introduced into Canadian politics an

[41] *Ibid.*, 3:5 (July 7, 1919).

[42] *Ibid.*, 3:5 (September 11, 1919).

[43] *Grain Growers' Guide*, September 1, 1920; *Manitoba Free Press*, December 3, 1920; *ibid.*, December 28, 1920.

[44] *Grain Growers' Guide*, 13:22 (September 1, 1920).

[45] *Ibid.*, September 15, 1920.

[46] Norman Lambert in the *Toronto Globe*, April 18, 1917, quoted by the Canadian Reconstruction Association pamphlet, "The Nonpartisan League in North Dakota."

aggressive group consciousness which profoundly influenced the development of the farmers' movement in Alberta and Saskatchewan, and to a lesser degree in Manitoba. Because the political life of the league was brief, the tendency quickly developed in Canada to look upon it as a flash of agrarian radicalism which burned fiercely but briefly, to die without a trace. The later successes of the United Farmers in politics obscured from many the fundamental contributions of the league to western Canadian agrarianism. If only because it was the first organization in Canada to use the recall and to elect a woman legislator, it deserves more than a footnote in Canadian agrarian history!

But it did more than that. The league was a harbinger of things to come in the political life of Canada. Its propaganda for political action indoctrinated thousands of Canadian farmers in the effectiveness of group action. Its ventures into the political arena anticipated the plans of George Chipman and the farm leaders associated with the *Grain Growers' Guide*. They were themselves looking forward to launching a farmers' party, and they saw in the league a dangerous rival. The bitter denunciations of the league were based on the knowledge that carefully prepared plans for a farmers' party would be destroyed if the league thrived. The *Guide* fought the league so bitterly that Chipman, on at least one occasion, altered the meaning of a reporter's story of a league convention to change a favorable account to a hostile review.[47]

The league served as a powerful agent to force the U.F.A. into politics in Alberta and did much to compel the S.G.G.A. to adopt an active interest in political affairs in Saskatchewan. Much of the program of the league was adopted by both organizations, and many Nonpartisans believed with good reason that the principles of the league survived though the society succumbed.

The league was a lesson in practical politics. To those who participated in the campaigns, the experience gave invaluable firsthand information in the operation of political institutions, and it was a revealing object lesson for those who observed. One of the primary reasons for the failure of farmers' movements in politics was inexperience. The league provided the training school for the

[47] This information was given to the author by the reporter in question.

agrarian leaders who were later called upon to exercise political control in the U.F.A. and the Progressive movement. It may well be that this experience helps explain the longevity of the farmers' government in Alberta, while in other regions in North America agrarian political control quickly collapsed. It is highly significant that every member of the executive committee of the Alberta league was also a member of the U.F.A., and that prominent leaders of the S.G.G.A. were also league officials.[48] Unwittingly, the league provided a reservoir of trained political leadership for the older farmers' movements when they took the plunge into politics.[49]

In a larger sense the league prepared the ground for the seeds of political revolt which flowered in the Progressive movement. Western farmers were reached by the gospel of political action and indoctrinated in the necessity of group solidarity. Agrarian leaders were trained in government and schooled in the stern realities of politics. An aggressive legislative program was formulated and publicized among the farmers. These were the contributions of the Nonpartisan League to the future. The Progressive movement thus inherited valuable assets without which it would have faced immeasurably greater, and perhaps insurmountable tasks of indoctrination, organization, and personnel training.

[48] *Alberta Non-Partisan,* 3:7 (February 12, 1919). P. L. Craigen was director of the S.G.G.A., Mrs. S. V. Haight was president of the W.S.G.G.A. for several years, and D. J. Sykes was a prominent member of the S.G.G.A. in District 14 for many years.

[49] Among these were D. H. Galbraith, John Glambeck, W. D. Trego, William Irvine, P. L. Craigen, D. F. Boissevain, Rice Sheppard, J. W. Leedy, and D. J. Sykes.

VII

The War Years

The conflagration that spread from Sarajevo caught the farmers of the Canadian West, along with almost all other people in the New World, completely unprepared for war. Years of isolationist and pacifist indoctrination had created an illusion of security and an optimism which the events of August 1914 rudely shattered. Western civilization faced the catastrophe of 1914 with shocked disbelief, and the farmers of the prairies shared the worldwide disillusionment.

The outbreak of war for a time created a unity on the prairies which had been missing during the preceding decade of agitation and discontent. The grain growers' organizations joined the universal condemnation of German aggression, and called upon the farmers to join in democracy's death struggle against the "Prussian military maniacs." The *Grain Growers' Guide* warned its farmer audience that British civilization was on trial, and the real question to be settled on the fields of blood in Europe was "whether British justice can be upheld by British might, whether democracy or autocracy shall triumph."[1] Spokesmen for the western farmers were united in their belief that Canadian honor, national liberty, and the Canadian soul were at stake in the struggle.

The *Guide*, however, avoided the war fever and the chauvinist hysteria that swept the world in the opening months of World War I. This moderate editorial policy was misunderstood by many patriots, and brought bitter charges of indifference—even of pro-Germanism.[2] The attitude of the farmers' journal was similar to the spirit shown by the democracies twenty-five years later when they again went to war with an aggressive Germany. To the *Guide* the war was less a crusade than a disagreeable job that had to be done.

[1] *Grain Growers' Guide*, 7:1 (August 12, 1914).
[2] *Ibid.*, 9:14 (August 16, 1916).

The search for causes that followed the first shock of war led to an analysis of prewar Europe by scholars and laymen everywhere. The conclusions reached by the western farmers were disquieting to conservative Canadians, for they reflected the years of indoctrination that had preceded the outbreak of hostilities. Agrarian leaders were convinced that the underlying causes for the war were economic. They thought the false economic system based on jealousy, distrust, and avarice, which western agrarians had condemned for a decade, was responsible for international as well as national strife.[3] The official statement issued by the Canadian Council of Agriculture explaining the "Farmers Platform of 1916" bluntly declared, "The present war is being fought out over a struggle for markets."[4] Moreover, it was widely believed that the Kaiser had plunged into war to crush the rising Socialist party and that the Russian Czar had followed a similar course to crush the peasants.

The unity which followed the outbreak of war soon vanished under the impact of tensions created by the war economy. American neutrality, moreover, exerted a strong pull and influenced many in western Canada to remain unenthusiastic about Canadian participation in a "European war."

The economic effects of the war were much the same in the hard spring wheat belt on both sides of the international boundary. They were intensified in Canada, however, because the war was not only much longer but relatively more costly to the dominion than to the republic, and also because the Canadian West had a more vulnerable economy. On both sides of the line, economic tensions produced by the war accelerated the loss of flexibility of a frontier economy and seriously limited the country's capacity for adjustment. Governmental failure to devise adequate price controls permitted an inflation that worked a hardship on the consumers of both nations. In the wheat belt the economic conditions which accompanied hostilities in Europe stimulated demands for government regulation of many of the economic functions in society, and undermined the wheat farmers' loyalty to the *laissez-faire*

[3] Presidential address of R. C. Henders at Brandon, January 10, 1917. Manitoba Grain Growers' Association *Yearbook*, 1917, 24.

[4] "The Farmers' Platform" (Winnipeg, 1917), 7.

doctrine. To a considerable degree this helps to explain the "social-service state" philosophy of the Nonpartisan League in the American Northwest and of the Progressive movement in the prairie provinces.

The tremendous expansion of production in the hard spring wheat belt was a striking feature of the war years. This reversed a trend on the Canadian prairies which had closely followed the declining world price of wheat, for the field crop acreage in 1914 was less than in any year since 1910.[5] Government officials appealed to patriotism and profit to meet the emergency demand for increased production of wheat and other foodstuffs. The war role of the farmer was described by enthusiastic officials as no less essential than that of the man in the trenches. The emphasis by Allied governments on bread as the "only diet which sufficed in isolation and was therefore indispensable"[6] placed the burden of production on the North American farmer, particularly after the Turks eliminated Russian wheat from Allied markets by closing the Dardanelles. Western farmers produced beyond the most sanguine hopes of Allied officials. The western Canadian acreage increased nearly 80 percent from 1913 to 1919, and Canadian wheat exports jumped from 194.9 million bushels in 1913 to 624.4 in 1920.[7]

This expansion was accomplished despite serious handicaps. A critical shortage of labor increased the cost of production and created a severe situation in Canadian agriculture by 1918. The higher price of machinery added to the costs of the hard-pressed farmer, and an inhospitable nature created production problems, for after the bumper crop of 1915 the acreage yields declined. In some areas drought reduced yields below the margin of profit and brought poverty amid the plenty of a war boom.[8] To step up production, farmers often purchased new lands at inflated prices, thus increasing the burden of debt carried by the West. Yet another cost

[5] *Canadian Annual Review of Public Affairs*, 1914, 259.

[6] *First Report of the Royal Commission on Wheat Supplies*, 1921, 9, quoted by Mitchell W. Sharp, "Allied Wheat Buying in Relationship to Canadian Marketing Policy, 1914–18," *Canadian Journal of Economics and Political Science*, 6:372 (August 1940).

[7] *Canada: Report of the Royal Commission on Dominion-Provincial Relations 1867–1939*, 1:90–92.

[8] The average yield in Alberta sank to six bushels an acre in 1918.

was added in 1917 when the Railway Commission allowed the railways to increase their freight rates by abrogating the Crow's Nest Pass agreement. The commission justified its action on the ground that the railways had to pay more for labor and materials and had to meet increased taxes. The Canadian Council of Agriculture vigorously opposed this increase as unjustified and as a menace to increased agricultural production.[9]

Agrarian leaders clearly saw the danger of an overexpanded wheat economy and early voiced their fears of a postwar collapse. The organized farmers' representatives, for example, refused to endorse an elaborate plan advanced by the Canadian Manufacturers Association during a convention of industrial and agricultural leaders in Winnipeg in 1914. The farmers maintained that instead of opening new lands with inexperienced farmers, it would be wiser to induce existing cultivators to increase their crop areas. The Regina delegation pointed out that, though they recognized the imperative necessity of increased production to win the war, the "bringing of larger areas under cultivation would be a step in the direction of increasing our wealth, only provided those who would engage in this work were able to farm the land at a profit to themselves."[10] Indeed, this profit must rest, the farmers believed, on a stable basis, and its continuity must be guaranteed to prevent a serious reaction following the war.

The *Guide*, recalling that government experts had urged "mixed farming" before the war to prevent a collapse of the economy, wondered if the emphasis on wheat production would not spell disaster for the prairies. W. R. Motherwell, one of the founders of the grain growers' movement and minister of agriculture for Saskatchewan, sharply criticized the dominion "Production and Thrift" campaign. On one occasion Motherwell claimed that if Canada failed to export a single bushel of wheat in 1916, it would be "scarcely missed in the world's general supply."[11] He often warned Saskatchewan farmers of the dangers of overexpansion and urged the agricultural community to "farm better but not more" if it would realize the fullest capacity. Agrarian leaders

[9] *Manitoba Free Press,* June 26, 1917; *ibid.,* June 29, 1917; *Grain Growers' Guide,* 11:5 (January 2, 1918).

[10] *Canadian Annual Review of Public Affairs,* 1914, 261.

[11] *Ibid.,* 1916, 694.

were fully aware of the dangers inherent in the war economy and sought to cushion the coming collapse in the world of wheat by preventing an abnormal increase in the number of producers. Though there were farmers who blindly reaped their war profits unmindful of the impending debacle, there were many more who feared future evils even more than they resented past grievances.

One of the complex economic problems that vexed western agriculture during the war was the question of fixed prices for wheat. Unfavorable marketing conditions had been a source of grievance for many years, and experience during the war added to the discontent. The Allied governments failed to develop a comprehensive purchasing policy during the first two years of the war, and the Canadian government resisted every effort to impose controls on the wheat economy until necessity forced action in 1917. When W. F. Cockshutt advocated price controls on food products in the Canadian House in 1915, Sir Robert Borden defended his government's policy by maintaining that the British market effectively controlled prices and that it was not desirable to reduce the farmer's price when the demand was strong.[12] The farmer, of course, also opposed such proposals, inspired as they were by a desire to reduce the cost of living at his expense.

The British wheat purchasing program was altered in October 1916 when the government abandoned its *laissez-faire* policy in favor of strict government control of cereal purchases. As a result, London approached the Canadian government in February with an offer to buy the whole surplus of the 1917 crop at the rate of $1.30 per bushel for No. 1 Northern in storage at Fort William.[13] Sir George Foster, the Canadian minister of trade and commerce, immediately sought the organized farmers' approval through their Canadian Council of Agriculture.[14]

The council rejected the offer as too low and countered with a proposal recommending $1.50 as a minimum and $1.90 as a maximum price. If a flat price were required, the council proposed $1.70 per bushel for No. 1 Northern as a fair return. This decision

[12] *Ibid.*, 1915, 268.

[13] Sharp, "Allied Wheat Buying in Relationship to Canadian Marketing Policy, 1914–1918," *Canadian Journal of Economics and Political Science*, 6:390 (August 1940).

[14] *Manitoba Free Press*, March 16, 1917.

was defended on the grounds that production costs had increased enormously during the war. It was estimated that it would cost the farmer at least $1.00 a bushel to get his wheat to the country elevators, and when freight and storage charges were added, the total cost to the farmer would be nearer $1.25 a bushel.[15] On top of this, the *Guide* argued that the farmers would be poor businessmen to accept a price lower than the average during the preceding six months.[16] The council maintained that its price proposals were designed to combine a maximum stimulus to production, with a minimum of opportunity for injurious speculation. With this advice, the government rejected the London offer.

Crises in the wheat markets of Winnipeg and Chicago eventually forced government control in both countries. Prices skyrocketed to $3.05 on the Winnipeg market as British purchasing agents tried to get all the wheat they could. With unlimited credit facilities, the Wheat Export Company purchased more contract grades than were actually available. Private companies ceased buying, for it was impossible under these conditions to protect their purchases by hedging. To prevent another such crisis, the Canadian government appointed a Board of Grain Supervisors which was to provide for an orderly marketing of future wheat crops and fix prices as it deemed necessary.[17]

The board proceeded to fix the price of the 1917 crop at $2.21 for No. 1 Northern in storage at Fort William. This was a liberal price, and was applauded by most grain growers as a 91-cent gain over the original offer rejected by the Canadian Council of Agriculture. There were dissenters, however, who believed that it was unfair to the farmer to set a price for his wheat and at the same time fail to control the prices of the manufactured articles he was forced to buy.[18] Then, too, there was always a group who believed that greater profits could be gained by allowing wheat to seek its "normal price level" in a free market. Nonetheless, the board continued its control over the 1918 crop, setting the price at $2.24½ per bushel.

[15] *Ibid.*, March 17, 1917; *Grain Growers' Guide*, 10:5 (March 21, 1917).
[16] *Ibid.*, 10:5 (March 21, 1917).
[17] *Ibid.*, 10:42 (May 19, 1917); Sharp, *op. cit.*
[18] Memorial to Sir Robert Borden from the executive of the Alberta Nonpartisan League, October 6, 1917, *Alberta Non-Partisan*, 1:7 (October 12, 1917).

The series of events that led to price fixing and the control of the markets by government intervention had a profound influence on agrarian thought on the prairies, and the farmers' reliance on a free market and their faith in the *laissez-faire* doctrine were further impaired. Precedents were created which later served as potent arguments when farmers faced declining prices in the postwar years.

The inflation that accompanied the war imposed burdens on Canadian as well as American farmers. Protests against the high cost of living were common in both countries. Canadian farmers, however, had little faith in the royal commissions appointed to investigate the high cost of living in the dominion. R. C. Henders, for example, minimized the work of such commissions as "superficial," for they dealt with effects rather than causes. The basic causes of inflation, he felt, were those which the farmers' organizations had long condemned: the "false system of land tenure," unearned increment, and an unfair system of taxation. The Manitoba president expressed a typical agrarian viewpoint when he maintained that taxation of land values, government ownership of public utilities, and free trade were the answers to the rising cost of living on the prairies.[19]

Farmers also expressed keen resentment against the "war profiteering" they professed to see on every hand. Though the grain growers' organizations were never as outspoken on this issue as the Nonpartisan League, they made it clear through numerous resolutions and memorials that they considered war profiteering vicious and unpatriotic.[20] The *Guide* believed that Canada's "worst enemies . . . are to be found right here in our own country." The farmers' journal not only condemned war profiteers as "patriots who bleed their country," but lashed out at mortgage companies and money lenders who extorted exorbitant rates from the farmers and foreclosed mortgages during the disasters of war and drought. The *Guide* warned that these were "the enemies that Canada has most reason to fear."[21] The high cost of living and war profiteering were frequent subjects of editorials and cartoons during the war.

[19] Manitoba Grain Growers' Association *Yearbook*, 1917, 27–28.

[20] *Grain Growers' Guide*, 9:7 (June 21, 1916); *ibid.*, 8:5 (March 24, 1915); *Canadian Annual Review of Public Affairs*, 1916, 723.

[21] *Grain Growers' Guide*, 7:5 (October 7, 1914).

A favorite theme for the *Guide's* cartoonists was a picture of John Bull valiantly fighting German militarism while unscrupulous Canadian war profiteers picked his pockets.

Western agrarians believed that inflation and war profiteering could be checked. They advocated a "conscription of wealth" program which relied on a graduated income tax to reduce profits. More drastic measures were sometimes proposed. The Saskatchewan grain growers resolved in their annual convention in 1916 that outright confiscation of earnings of more than 15 percent in munitions should be utilized to prevent undue profiteering.[22] Agitation against war profits and inflation grew in intensity among northern wheat farmers and increased the discontent on both sides of the forty-ninth parallel. The radical elements joined the Nonpartisan League in both countries, and even the more cautious were beginning to doubt that a solution could be found within traditional parties and through orthodox measures. Few farmers, however, associated their increased purchases of land, new machinery, and automobiles with the war profiteering they saw in the cities.

The banking and monetary systems in western Canada came under sharp criticism during the war years. The drying up of the flow of credit that followed the financial crisis of 1913 was so intensified by the war that farmers often found it impossible to secure credit accommodation through the usual channels. Vere Brown, superintendent of the central western branches of the Canadian Bank of Commerce, often decried the financial burdens carried by the agricultural community. His analysis, the result of an intimate knowledge of the Canadian financial structure and of western agriculture, found four major evils in the existing system: excessive mortgage interest, excessive bank interest, excessive costs for goods due to a "bad credit system," and the protective tariff.[23]

Other businessmen were equally critical of the banking system in the West. Western retail lumbermen frequently complained of the heavy credits they were forced to extend to farmers over long periods. The bitter experience of the war years led the lumbermen's association to favor "changes in the Canadian banking sys-

[22] *Canadian Annual Review of Public Affairs,* 1916, 723.
[23] *Manitoba Free Press,* March 7, 1917.

tem so as to take care of the needs of the farming industry that it will not be necessary for the lumbermen to extend long credits."[24]

Premier A. L. Sifton of Alberta was keenly aware of the financial problems of his province, and he frequently advocated measures to secure "cheaper money" for the farmers and "a more definite control over deposits" by the people. His suggestion to incorporate hundreds of small banks throughout the West was a familiar western solution for a complicated problem.[25] Similar schemes had been advocated in frontier regions since earliest colonial days in North America.

Providing "cheaper money" through provincial agricultural banks was the object of numerous resolutions by the grain growers' organizations,[26] and the Canadian Council of Agriculture held conferences with banking representatives to secure better terms for essential short-term loans. Farmers claimed, with considerable justice, that the Canadian banking system had been created in eastern Canada for a commercial economy. The transposition of this system to a frontier agricultural economy worked a hardship on producers and retarded the development of the country.

Agitation for reform of the financial system was especially strong in Alberta, where the pioneer economy, already suffering from credit malnutrition, was further pinched by drought and war. The membership of the U.F.A. included numerous ex-Americans whose experience with financial and monetary questions was extensive. Agrarian organizations in the states had emphasized financial reform for many years, and the ex-Americans were thoroughly indoctrinated in these issues. Many of these men, moreover, were experts in agrarian finance and had helped frame legislation in the states from which they had emigrated.

J. W. Leedy's papers on credit and monetary reform, for example, excited much attention.[27] Leedy possessed a keen understanding of financial questions; he had authored financial legislation while he was Populist governor of Kansas in the 1890's. Obviously inspired by the brilliant essay of a more famous fellow Kansan,

[24] Resolution of the Western Retail Lumbermen's Association, Calgary, Alberta, January 12, 1919 in *Lumber World Review* (Chicago), 36:38 (February 1919).

[25] *Canadian Annual Review of Public Affairs*, 1916, 728; *ibid.*, 1915, 690.

[26] *Ibid.*, 1915, 708; *ibid.*, 1916, 653, 746.

[27] For an example see the *Grain Growers' Guide*, 9:5 (January 26, 1916).

William Allen White, Leedy published a pamphlet, "What's the Matter with Canada?" which discussed these issues. George Bevington, who was to play a more conspicuous role in the postwar years, had studied monetary questions for many years and was ably seconded by other U.F.A. leaders, especially Robert Gardiner and Henry Spencer. In a very real sense, the vigorous agitation of these men for monetary reform prepared the way for the "funny money" doctrines which swept the prairies during the Great Depression.

Agitation for credit reform during the war years led to the organization of cooperative credit agencies and produced provincial legislation to ease the situation. In 1917, for example, both Manitoba and Alberta took action in the form of a rural credits act and a farm loan act. These acts sought to reduce interest rates by lending money at cost to the farmers. Agricultural development and increased production would be stimulated, it was hoped, by reducing the collateral necessary to secure a loan and by equalizing interest rates among the agencies offering rural credits. Discussion and criticism of financial and monetary conditions thus stimulated remedial legislation and anticipated much that was to happen in later years on the prairies. Many farmers had wearied of the fight against the protective tariff and sought economic relief through credit and currency manipulation.

These complaints should not be allowed to conceal the fact that the wheat economy of the West assumed an unprecedented importance during the war. The urgency of increased wheat production to prevent widespread suffering among the Allied peoples of Europe exaggerated the importance of the prairies in the Canadian economy. This temporary economic position in turn magnified the political importance of the prairie provinces in dominion politics. This political weight, which persisted into the early years of the postwar period, did much to shape the destiny of the dominion during the early twenties, and is one of the explanations for the rise and fall of agrarian political strength from 1914 to 1926.

Because of the strain of the war it was impossible to lay aside the agitation for reform that had grown so strong in the prewar years. Responsible agrarian leaders were well aware of the danger of continuing the crusade for reform. They realized that criticism of existing institutions created fissures in the social structure which were especially dangerous during the crisis of war. Henry Wise Wood urged the members of the U.F.A. to engage in the work of production and cease quibbling over alleged wrongs and grievances. There was real danger, he warned, that the farmers might emerge from the war "a discredited class with none to do them reverence." [28] The warnings of cooler heads were largely ignored, however, for the habits of a decade could not be changed so quickly.

There were many evidences during the war of a deep and abiding discontent on the prairies. The increasingly radical tone of western liberalism, the rapid expansion of the grain growers' associations and the tremendously increased circulation of the *Grain Growers' Guide*,[29] the appeal of the Nonpartisan League in Saskatchewan and Alberta, and the growth of radical journals such as the *Nutcracker* were striking indications that prewar grievances had not been eliminated by world conflict. The agrarian revolt that swept North Dakota in 1916 was the shadow of coming events in the Canadian West.

The discontent inherited from the prewar decade was strengthened by events of the war years. Numerous political scandals confirmed the farmers' conviction that regular parties were hopelessly corrupt and intensified their resentment against politicians.

The governments of all three prairie provinces came under suspicion during the war. The powerful Roblin-Rogers machine in Manitoba fell apart under the pressure of the legislative building scandal in 1915 and left Conservatives in this oldest of the prairie provinces dispirited and dejected. The Walter Scott government in Saskatchewan faced serious charges of graft and bribery during the following year. Three royal commissions investigated charges of fraud in liquor licenses, highway funds, and provincial institutions.[30] When Premier Scott retired under the plea of broken health, the farmers' control of the Saskatchewan Liberal party was

[28] *Manitoba Free Press*, January 30, 1918.
[29] Membership in the associations had grown to approximately one hundred thousand by October 1918, and the circulation of the *Guide* passed the forty thousand mark in that year.
[30] *Canadian Annual Review of Public Affairs*, 1916, 704–6.

strengthened with the naming of William M. Martin as premier, and the able Charles A. Dunning as provincial treasurer.[31]

Charges of corruption were also leveled at the Liberal government in Alberta. Premier A. L. Sifton was hard pressed to defend his administration against accusations of fraud in the handling of liquor licenses. The resignation and suicide of the offending minister more than a year later seemed to substantiate many of the charges. Moreover, rumors of graft and corruption in the dominion government circulated freely on the prairies. Scandals in the purchase of horses, binoculars, ammunition, rifles, and boots were cited by the *Guide*, the *Nutcracker*, and other western journals as evidence of the moral bankruptcy of "partyism."

The Farmers' Platform of 1916 was the most striking move in the agitation for reform during the war years. Drafted by the Canadian Council of Agriculture from the hundreds of petitions, resolutions, and memorials received from grain growers' locals during the preceding years, it defined the future course of agrarian demands. Early in 1917 the platform was submitted for approval to the annual conventions of the grain growers of the West and the United Farmers of Ontario. Having secured their approval, the council then submitted the platform to each of the two thousand local unions, which rallied behind it. The farmers were being well educated for a political revolt.

The Farmers' Platform introduced no new demands. It summarized clearly and in unmistakable terms the farmers' grievances, particularly those against the protective tariff. The document called for all the free-trade measures that had been demanded for many years. It called attention to the fact that the Reciprocity Agreement of 1911 remained on American statute books and asked that the offer be accepted. Taxation for revenue through the imposition of a direct tax on unimproved lands and natural resources, sharply graduated personal income and inheritance taxes, and a graduated tax on corporation profits were urged as a solution to the fiscal problem. These demands had been the core of agrarian agitation for nearly twenty years, and they occupied 750 words of the 950-word platform.

[31] Dunning at the time was general manager of the Saskatchewan Cooperative Elevator Company, vice-president of the S.G.G.A., and a member of the Canadian Council of Agriculture.

The remainder of the document endorsed other familiar reforms such as direct legislation, abolition of political patronage, the nationalization of all railway, telegraph, and express companies, and woman suffrage.[32] These, too, were the old cries that had been heard on the prairies for many years. The platform presented to the Canadian people an authoritative expression of the grain growers' program. Thus it charted the course of future political action and served as a rallying point for the farmers and their friends in the coming struggle for power.

The demand for the nationalization of railways was pointed and timely. Few problems were as oppressive to the Canadian government during the war as the railway situation. Finding a solution to the railway problem was second only to the conduct of the war, and at times it occupied the Borden ministry almost exclusively.[33] The financial distress of the two newer transcontinental systems, the Grand Trunk Pacific and the Canadian Northern Railway, became increasingly acute under the stress of war. By 1916 it was quite clear that these companies could no longer carry on, because it was impossible for the government to continue annual subsidies to them and at the same time meet the strain imposed on Canadian resources by the war.

The position of the western farmers in the controversy was perfectly clear. They bitterly condemned government subsidies to the railways, especially to the Mackenzie-Mann interests, which the farmers were convinced had milked the government for years and had corrupted the Conservatives with ample donations to their party slush fund.[34] The railway problem became so serious that the government was forced to appoint a commission to investigate possible solutions. Acting upon the Drayton-Acworth Report, the government moved to nationalize the distressed railways. This step was generally approved by the grain growers, except that it did not go far enough. They wanted it to include the Canadian Pacific: "If the people of Canada," complained the *Guide*, "are to become proprietors of the two lean railways there is all the more reason why they should take over the fat one at the same time."[35]

[32] "The Farmers' Platform" (Winnipeg, 1917), 5.
[33] Henry Borden, ed., *Robert Laird Borden: His Memoirs*, 2:643–55.
[34] *Grain Growers' Guide*, 4:5 (May 6, 1914); *ibid.*, 7:5 (March 25, 1914).
[35] *Ibid.*, 10:5 (May 9, 1917).

The grain growers also renewed their emphasis on the need for social reforms and were successful in securing important legislation during the war years. The farmers' organizations claimed credit for the wartime prohibition acts, woman suffrage, compulsory education, and direct legislation measures enacted by provincial legislatures. The opposition to the "institution of a blooded and titled aristocracy" gained momentum, and public opinion in the West clearly condemned the "tin-pot titles."[36] The *Guide* believed that the whole system had brought titles into disrepute among the rank and file of Canadians.[37] In March 1918 the Borden ministry, recognizing the strength of this sentiment, issued an order-in-council that declared hereditary titles incompatible with Canadian ideals of democracy.[38] The British secretary of state for the colonies was advised that no hereditary title should be conferred upon a "subject of His Majesty ordinarily resident in Canada." Sir Wilfrid Laurier, then leader of the opposition, approved this action, called for the abolition of all titles, and dramatically announced that he was prepared "to bring all our titles to the market place and make a bonfire of them."[39] To all this, the grain growers of the West said amen!

The farmers' organizations wrestled, from the earliest days of the war, with the problem of writing a just peace. The quest for social justice, which was so much a part of the grain growers' movement, made this inevitable. In its first wartime convention the U.F.A. discussed the nature of the coming peace, and expressed its conviction that the chief aim of the peace terms must be to prevent future wars. To effect this the farmers called for a peace treaty which avoided "conquest and national humiliation" and satisfied "all legitimate national aspirations."[40] This counsel of perfection was typical of much western thinking and reflected the idealism of the grain growers' indoctrination. They denounced militarism and imperialism and called for complete disarmament after the war.

The farmers of the West threshed out the issues involved in the

[36] *Nutcracker,* 1:13 (March 2, 1917).
[37] *Grain Growers' Guide,* 11:5 (April 3, 1918).
[38] Henry Borden, ed., *op. cit.,* 2:794.
[39] House of Commons *Debates,* Session 1918, 1:500.
[40] *Canadian Annual Review of Public Affairs,* 1915, 705.

writing of a peace and as a class were probably as well informed on these questions as any comparable group in North America. R. C. Henders expressed the feeling of many grain growers when he called for an international court clothed with power to enforce its decrees and an international police force made up of air, land, and sea units.[41]

The grain growers' reactions to international events are revealing. American entry into the war was greeted with enthusiasm, and Wilson's announced war aims were heralded as those of the farmers' organizations. "Never since the days of Abraham Lincoln," thought the editor of the *Guide*, "has any man in a responsible position been called upon to make such a momentous decision."[42] There was little criticism in western Canada of America's long delay in entering the struggle—perhaps because Canadian farmers remembered their own prewar isolationism and because the heterogeneous character of prairie population gave western Canadians an understanding of the problem of devising a unified foreign policy in the face of large, partially assimilated national minorities.

Farmer reactions to the early stages of the revolution in Russia were those of liberals everywhere. The grain growers watched the overthrow of the Czar's government with sympathy and believed that democracy had at last reached eastern Europe. The *Guide* expressed this wide-felt sympathy for Russian democrats and believed the revolution would strengthen the positions of those on the American continent who were struggling for greater political freedom.[43]

The grain growers were concerned about any trends in North America which challenged their concepts of democracy. The *Grain Growers' Guide*, for example, was deeply interested in academic freedom in Canadian and American colleges, and it followed with keen interest the efforts to oust President Ladd and Professor Bolley of North Dakota State College, whose experiments with wheat flour furnished evidence to the Equity and the Nonpartisan League in their struggle with the grain dealers. When the Reverend Dr. Salem G. Bland was summarily dismissed from Wesley College

[41] *Ibid.*, 1916, 652.
[42] *Grain Growers' Guide*, 10:5 (April 11, 1917).
[43] *Ibid.*, 10:5 (March 28, 1917).

after fourteen years of service, the *Guide* professed to see the same insidious forces at work in Winnipeg as in Fargo. Dr. Bland was defended by the farmers' paper, which believed the real reason for his dismissal was an undue friendliness to the grain growers and to other liberal elements in western Canada. The two incidents were cited as stern warnings that democracy must always be on the alert "to prevent the muzzling of our educational institutions."[44]

Canadian farmers were severely criticized during the war for the same indifference, selfishness, and war profiteering they had condemned in others. The agricultural community was accused of complacency in the face of Canada's struggle for national existence. Agriculture, many believed, was not carrying its fair share of the load, and some hinted that the enlistment of farmers' sons lagged far below that in the more patriotic urban areas. Farmers were censured for failing to support wartime relief agencies, and they were rebuked as "more concerned with the rights and interests of the farmers" than they were with the war effort.

The war role of agriculture was misunderstood by many who, seeing the dramatic transfer of manpower into uniform and into war industries, damned the farmer for staying on his land to carry on a routine that appeared the same as his peacetime schedule. Stephen Leacock, in a propaganda pamphlet distributed by the government, proved that he was more a humorist than an agricultural economist when he disapproved the idea of "teaching the farmer that his own fat, easy industry is war itself, and that he may count his fatted cattle in the light of his stable lantern and go to bed a patriot."[45]

Superficial analysis condemned the farmer and suspected sedition in his continued agitation for reform. Contributions through the Patriotic Acre Fund, the Red Cross, and the Belgian, Armenian, and Polish relief funds were often overlooked, as was the fundamental importance of prairie wheat to Allied victory. The United Farmers of Alberta were frequently the target of particularly bitter criticism. The *Manitoba Free Press* professed to see in

[44] *Ibid.*, 10:6 (July 18, 1917).

[45] Stephen Leacock, "Our Organization for the War," reprinted in J. O. Miller, ed., *The New Era in Canada*.

the organization a "spirit of indifference and detachment which was alarming" and criticized Henry Wise Wood, the organization's president, for being more concerned with the interests of the farmers than with how "our freedom was to be preserved."[46] The high percentage of ex-Americans and foreign-born in its membership made the society extremely vulnerable. The *Free Press* deplored the tendency of U.F.A. members to cite American experience "rather than what is done in Ottawa," and suggested that these people be made "Canadian and British in spirit as well as in name."[47]

The *Guide* answered these charges by maintaining that "if noise, flag-flapping, and resolutions are true measures of loyalty, patriotism, and Canadian sentiment," the U.F.A. was open to criticism. The journal also defended the ex-Americans, claiming that no better citizens had come into the country and that the U.F.A. members represented the "highest type of patriotism and public spirit, and are in accord with the Golden Rule."[48]

Fundamentally, of course, Canadian agrarians were in danger of falling into the same situation that discredited the Nonpartisan League in North Dakota and led A. C. Townley into the courts on charges of sedition. Reform and war are not happy partners. Years of critical analysis of the economic system and of Canadian society had left the farmers ill prepared to join the hysteria and chauvinism of World War I. Criticism leveled at Alberta farmers, moreover, overlooked the fact that this western province actually did more than its share in providing men for the army and in 1916 contributed a higher percentage of eligible men than any other province. The *Edmonton Bulletin* proudly and properly felt that "the call has been answered in a manner that must make every heart thrill with pride."[49]

Sentiment for political action grew rapidly during the war. The agitation for an outright break with the regular parties was supported by thousands of western farmers who condemned their leaders for excessive timidity in the face of urgent needs.

[46] *Manitoba Free Press*, January 30, 1918; *ibid.*, February 7, 1917.
[47] *Ibid.*, January 30, 1918.
[48] *Crain Growers' Guide*, 10:5 (February 14, 1917).
[49] House of Commons *Debates*, Session 1919, 1:258; *Canadian Annual Review of Public Affairs*, 1916, 743; *Edmonton Bulletin*, October 18, 1915.

The urge to action came from below more than from above. As already indicated, some of the leaders were bitterly opposed to an entry into the political arena. Henry Wise Wood was a vigorous opponent of an independent farmers' party, his own experience with the alliance in Missouri having taught him the lesson of caution. Wood warned that disaster could easily follow a political venture. The farmers, he argued, were unskilled in politics and their entrance into elections would attract "all the discredited, broken-down politicians of both parties."[50] The old spirit of party prejudice could easily break the farmers' unity, cripple the powerful cooperative movement, and undermine the class organization that grain growers had built through the years. But in spite of these and similar warnings from other leaders, the rank and file were determined to take the plunge into politics, and their influence in the provincial elections of 1917 in Saskatchewan and Alberta was a clear indication of the strength of this feeling.

Agitation for a "free-trade party" reached its wartime peak early in 1917. The probability of a federal election before harvest stirred excited speculations,[51] and demands for an independent party were heard on every hand. Editorials, cartoons, and feature articles in the *Guide* fed the growing appetite for politics. "Forget the old parties," urged the editor, but he warned against using the machinery of the established farmers' organizations. Earlier he had advised local farmers to find good candidates who were honest democrats, in touch with the work of the organized farmers, and possessed of the courage to stand for their principles.[52]

There were hopes that a political reformation could be effected through the historic Liberal party by capturing the organization and by transforming it into a genuinely liberal party. "Political action is necessary and can be carried on without interference or danger to the farmers' organizations," the *Guide* advised its readers in 1916.[53] Charles A. Dunning of Saskatchewan strongly supported this course. "I believe," he once declared, "that from the West . . . it is possible to develop such a force for true Lib-

[50] Henry Wise Wood, "The Organized Farmers and Politics," *Grain Growers' Guide*, 10:10 (September 19, 1917).
[51] *Ibid.*, 10:5 (March 7, 1917).
[52] *Ibid.*, 8:5 (April 21, 1915).
[53] *Ibid.*, 9:5 (January 19, 1916).

eralism as will affect the whole of the party, and so render reform easy."[54]

There were many indications that such might be the course of political events on the prairies. The "Bill of Rights" adopted by the Saskatchewan Liberal convention in April 1917 was strikingly similar to the Farmers' Platform, and the *Manitoba Free Press* welcomed it as evidence of a new and virile type of liberalism that would redeem Canada if eastern Liberals would accept it.[55] This was the catch: Western radicals were blocked in their efforts to control the party by protectionist eastern Liberals who were Liberal in name but conservative in spirit.

The *Free Press*, under the skilled hand of John W. Dafoe, encouraged the trend. "The time is ripe," declared this organ of prairie Liberalism, "for Western Liberals to decide that they will rely upon themselves — and thus do their own thinking, formulate their own policies, and provide their own leaders. Canadian public life will thus be given what it sorely needs — a group of convinced radicals who will be far more interested in the furthering of their programme than in office holding."[56] Agrarian writers joined the movement to form a radical wing of liberalism on the prairies. They pointed to the success of British radicals in capturing the Liberal party in England and believed that western agrarians could duplicate the feat in Canada.[57] Before such plans could mature, however, the bitter controversy over conscription and the "Win-the-war" campaign intervened to split the Liberal party into two factions, conscriptionists and anti-conscriptionists, thus eliminating it as an effective agency of the farmers' will.

The conscription issue was disruptive in the extreme. The division of the Liberals into hostile factions greatly weakened the party in the West, and shaped the course of postwar dominion politics by leaving a political vacuum into which the farmers eagerly rushed with their Progressive movement.

Conscription became a public issue in May 1917 when the Borden government felt forced by events in Europe to adopt it.

[54] *Canadian Annual Review of Public Affairs*, 1916, 722.
[55] *Manitoba Free Press*, April 3, 1917.
[56] *Ibid.*, May 13, 1916.
[57] J. A. Stevenson, "Future of Western Democracy," *Grain Growers' Guide*, 9:8, 24 (July 12, 1916).

Though western farmers generally supported the move, they did so with the reservation that farm labor would not be depleted below its then critical level. Many farmers also urged the conscription of wealth along with that of manpower.[58] The United Farmers of Ontario called for a referendum to let the people decide the issue as they had done in Australia. Western grain growers went farther than this and urged the government to hold a referendum on conscription, not only of men but also of wealth.

Canadians were soon called upon to approve conscription in an election that came close to being the referendum demanded by the farmers. Faced by the serious position of the Allied armies in Europe, and seeing an opportunity to crush the Liberals and secure a united political front, Sir Robert Borden reorganized his ministry in August 1917. Though Sir Wilfrid Laurier refused to enter the cabinet, the prime minister found sufficient conscriptionist Liberals who were willing to forsake party for country during this national crisis. The organized farmers were also consulted during the negotiations. Both T. A. Crerar and H. W. Wood journeyed to Ottawa for conferences with the prime minister. Wood, however, refused to be a candidate for a cabinet position.[59]

The major political problem facing the Borden ministry was to present conscription to the people through a coalition government by detaching conscriptionist Liberals from their federal party.[60] An attempt was made to accomplish this in a western Liberal convention in Winnipeg in August 1917. "Regular Laurier" delegates, however, captured control of the convention and prevented the planned separation.[61] This maneuver frustrated western Liberal conscriptionists only temporarily. They proceeded to accomplish their aims by direct negotiation with the prime minister. The discussions covered a period of twenty weeks and resulted in the

[58] *Ibid.*, 10:5 (March 23, 1917); *Nutcracker*, 1:9 (February 3, 1917).

[59] John W. Dafoe, *Clifford Sifton in Relation to His Times*, 425. Arthur Ford, an observer during the negotiations, claimed that "Mr. Wood, who was American-born and not familiar with Canadian politics, Canadian history or Eastern Canada, refused." ("Notes on Formation of Union Government, 1917," *Canadian Historical Review*, 19:357 [December 1938].) H. W. Wood's complete lack of political ambition or desire to assume political responsibility was probably also a factor. Wood later admitted that political office seemed "revolting rather than attractive." ("Wood Papers," manuscript in the possession of L. D. Nesbitt.)

[60] W. L. Morton, "The Western Progressive Movement, 1917–1921," *Proceedings* of the Canadian Historical Association, 1946, 44–45.

[61] *Grain Growers' Guide*, 10:3 (August 15, 1917).

entrance of nine prominent Liberals into the cabinet, including T. A. Crerar, A. L. Sifton, and J. A. Calder from the West.[62]

The new union government appealed to the people in a general election in December 1917. Though the farmers had long fought for a united, nonpartisan war government, they supported "King Borden" with manifest reluctance. The *Guide* eventually threw its influence to the union candidates, but with the warning that "no milk or water fusionist candidate nor anything of the nature of a rubber stamp 'Win-the-war' candidate can truly represent the spirit of the Western people."[63] The *Guide* had an eye to the future, for it was obvious that the union government would determine economic questions not directly related to the war effort. The farmers were willing to agree to a truce in the battle for free trade, but they wanted their representatives in Ottawa to feel free to resume the fight for the Farmers' Platform with the return of peace.

Five prairie constituencies elected "independent farmer candidates" to guarantee that postwar problems would be in reliable hands. Though each of these five men announced his independence of formal organizations and ran on the "Win-the-war" ticket, he held an official position in the grain growers' movement.[64] Their elections were victories for the grain growers and clear warnings of what was to happen later in all rural constituencies in the West.

The War Times Election Act, which enfranchised women relatives of soldiers and disenfranchised conscientious objectors and recently arrived persons of alien birth, made certain the victory of the union ticket but aroused keen resentment among western farmers.[65] The *Guide*, the *Alberta Non-Partisan*, and grain growers' locals condemned the act as a violation of "British justice" and as an example of the party politician's methods.[66]

The conscription issue was again raised in April 1918, when the critical Allied situation in Europe forced the union govern-

[62] Henry Borden, ed., *op. cit.*, 2:735–57.

[63] *Grain Growers' Guide*, 10:5 (October 17, 1917).

[64] The five men were T. A. Crerar, Marquette (president, United Grain Growers, Limited); R. C. Henders, Macdonald (president, M.G.G.A.); John A. Maharg, Maple Creek (president, S.G.G.A.); John F. Reid, Mackenzie (director, S.G.G.A.); Andrew Knox, Prince Albert (director, S.G.G.A.).

[65] House of Commons *Debates*, Session 1918, 2:1459.

[66] A resolution passed in District 14 of the S.G.G.A. is typical. It denounced the act as an example of the methods of "party politicians who have disenfranchised many British citizens under the guise of patriotism."

ment to extend the draft to large numbers of farmers' sons. Canadian farmers naturally looked upon this as a breach of faith, for the government had promised exemptions to farmers' sons in the 1917 campaign. They complained bitterly that further conscription of farm labor would force thousands of acres out of production. Western farmers were generally sympathetic with the mass meeting in Toronto held by the United Farmers of Ontario which denounced the Military Service Act and condemned the government which sought to enforce it. The *Alberta Non-Partisan* considered the excitement over conscription "highly amusing" and unkindly pointed out that "it is only a few short months since these farmers rushed to the polls and endorsed the very act they now protest against."[67]

Canadian agriculture, it is true, was suffering from a shortage of labor. Unfortunately, however, the farmers could not be told that mutinies in the French army in 1917 had threatened to paralyze the Allies and had forced the British to accept commitments which drained manpower reserves faster than they could be replaced.[68] The U.F.A. warned the government that the situation was "so serious it calls for the greatest possible wisdom . . . to maintain that hearty support that a loyal people owe to their government."[69]

The farmers' protests against the government's new conscription policy grew so vehement and unrestrained that many feared the farmers would discredit themselves before the Canadian people. Responsible agricultural leaders warned that cooler heads must not be carried away by the agitation against the government. The conscription issue unquestionably created an opposition to the union government that forever discredited it in the farmers' eyes. The cool, and some thought insulting, manner in which the government received the U.F.O. delegation in June added to the resentment. The farmers of Canada were determined that such a government must not be allowed to last.

During the last months of the war, western farmers anticipated

[67] *Alberta Non-Partisan,* 2:13 (May 24, 1918).

[68] Other military factors that forced greater manpower levies were the collapse of the Russian front, the successes of the German submarine offensive, and the decline of Italian effectiveness.

[69] *Grain Growers' Guide,* 11:25 (June 26, 1918).

with increasing interest the postwar struggle for political power.[70] A campaign was launched by the *Grain Growers' Guide* to prepare the way for the coming farmers' party, for there was no longer any doubt that it would appear. In the West the conscription issue had eliminated the Liberal party as effectively as the tariff question had wrecked the Conservatives. The "Win-the-war" campaign had destroyed traditional loyalties to the historic parties. Many farmers had broken the political habits of a lifetime, and federal parties were in the process of dissolution.

The union government had temporarily checked the growth of third-party sentiment, but only temporarily, for tensions had been increased by the war and the ferment that had worked on the prairies for nearly two decades had continued to grow beneath the surface. As J. A. Maharg warned in the House of Commons in May 1918, "We buried our economic views entirely last December. We have not changed our views at all. . . . In fact, war conditions have intensified them, and we are now more convinced that the views held by Western Canada for the last few years, have been economically sound."[71]

Thus when peace came in late 1918, the stage was set for political revolt in western Canada. The principal actors had rehearsed their parts for many years, and the voice of the organized farmers, the *Grain Growers' Guide*, was ready in the wings to prompt them should they forget their lines. The elections of 1917 had lifted the curtain slightly to give a preview of the drama that was soon to absorb the attention of the entire dominion audience.

[70] *Ibid.*, 11:10 (May 22, 1918).
[71] House of Commons *Debates*, Session 1918, 2:1459.

VIII

Revolt on the Prairies

The war at long last was over, but peace seemed to betray those who had sacrificed the most, and the postwar years brought disillusionment, restlessness, and discontent. The Great Crusade had ended victoriously, but victory failed to bring the promised "new world" that had called forth unstinted devotion and prolonged sacrifice. The Canadian West was infected with the same restless spirit that swept the world during the early twenties.

Canada faced tremendous tasks of demobilization and reconstruction when the guns were finally silenced on the western front. These major problems were aggravated by the economic collapse and political ferment that followed the return to peace. Postwar disillusionment was accompanied by industrial strife, political confusion, and international discord. Inflation, unemployment, and a staggering war debt imposed a burden on the average citizen which caused widespread pessimism. Many of the established political landmarks had been swept away during the war, and political parties were adrift as politicians found it hard to get their bearings in the new, uncharted seas of the postwar world. Sectional cleavages were revived with increased bitterness, and for the first time class conflict appeared in the dominion.

The severe economic dislocations that followed on the heels of peace caused deep resentment in the dominion, as they did in the United States. Bitter feeling against "war profiteering" was increased by the pinch of inflation and unemployment. The two-hundred-dollar per capita war debt pressed heavily upon the dominion.[1] On top of this, the West carried a heavy debt that had accumulated in the decade of expansion before the war. The serious railway problem remained unsolved as the anticipated flood of immigrants failed to materialize, and the overexpanded systems

[1] House of Commons *Debates*, Session 1919, 1:19.

were forced to rely upon the thinly scattered prairie population for revenue. The economy of the prairie provinces staggered under the burden of depressed agricultural prices, heavy fixed-debt charges, and a high level of production costs.

There were many indications of widespread discontent on the prairie. The hysterical Red scare that swept the Allied countries after 1917 found lodgment on the prairies where labor strife gave an air of reality to the fears. The high cost of living encouraged support for the agitation of International Workers of the World and led to the formation of the One Big Union in the Canadian West. Labor strife reached a dramatic climax in the tragic Winnipeg general strike in 1919, which seemed to confirm the fear of many farmers that Bolshevism was now firmly established in the cities.

The curious movement of the Ku Klux Klan into western Canada in the years following 1918 was another indication of the temper of the times. The Klan invaded Saskatchewan when professional organizers from the parent society in the Middle West quietly preached their gospel of nativism, pure Anglo-Saxonism, and religious bigotry. The *Farmers' Advocate* saw the movement as a serious menace to prairie society and warned its readers that the Klan "draws to its banner all kinds of crooks, law breakers and degenerates." The *Advocate* advised that the Klan "ought to be stamped on as one would stamp on a rattlesnake—and wherever it rears its ugly head."[2] In spite of clear warnings as to the true nature of the organization, the Klan continued its malignant growth until, in the opinion of one contemporary observer, the movement was taken so seriously on the prairies "as to be regarded as part and parcel of the social and political scheme of things."[3]

The Klan, the One Big Union, the "Wobblies," and the Red scare were symptomatic of the postwar hysteria and restlessness. They placed in jeopardy the farmers' crusade for reform, for there was now the danger that every liberal sentiment would be branded with the Bolshevist stigma.

The agricultural depression, which had been foreshadowed as

[2] *Farmers' Advocate*, 57:1181 (October 19, 1921).

[3] Anonymous, "Ku Klux Klan in Saskatchewan," *Queens Quarterly*, 35:592 (Autumn 1928).

early as 1909 and had been greatly feared during the war years, struck in mid-1920, bringing widespread distress in its wake. Farmers on both sides of the forty-ninth parallel suffered short crops, ruinous prices, and high costs of production.[4] Wheat quotations on the Winnipeg market plunged from $2.78 a bushel for No. 1 Northern in September 1920 to $1.76 in April 1921, and Minneapolis prices fell from $2.49 to $1.31 during the same period.[5] This price collapse was severe in itself, but since it was not accompanied by a similar drop in living costs, wages, freight rates, or interest charges, it worked a particular hardship in the agricultural community. Economic conditions in the hard spring wheat belt, as a result, were worse than they had been for thirty years.[6]

By 1923 the farmer's purchasing power had shrunk to less than 20 percent of its prewar level as his income was pinched between an increase in land values, labor costs, taxation, transportation costs, and a serious restriction of credit on the one hand, and short crops and ruinously low prices on the other. The unreasonably high costs of distribution and the economic depression in Europe deepened his distress.[7] The immediate postwar years witnessed renewed efforts by North American farmers to reduce the disparity in prices, to cut production and marketing costs, and to secure easier credit accommodation. They sought these objectives through expanded cooperatives, wheat boards, and political action. The high production costs and disastrous prices of 1920, which brought losses to thousands of farmers,[8] sharpened agrarian demands for reform and intensified grievances inherent in commercialized agriculture. Grievances of twenty years' standing were suddenly magnified, and they served as the immediate stimulant of the agrarian revolt in Canada and the formation of the powerful farm bloc in the United States.

The agrarian crusade in western Canada was dual in nature. The primary emphasis of the farmers' organizations was on eco-

[4] United States Department of Agriculture *Yearbook*, 1921, 56; *Canadian Annual Review of Public Affairs*, 1920, 93.

[5] *House Report* No. 408, 67th Congress, 1st Session, vol. 3, pt. I, "The Agricultural Crisis and Its Causes," 96.

[6] *House Report* No. 408, 67th Congress, 1st Session, vol. 3, pt. 2, 7.

[7] United States Department of Agriculture *Yearbook*, 1922, 3.

[8] *Ibid.*, 1921, 7; *ibid.*, 1922, 551; *Report of Royal Commission on Dominion-Provincial Relations 1867–1939*, 1:113.

nomic reforms, though political action attracted wider attention. In both countries during the postwar years, cooperative societies expanded rapidly as farmers sought to reduce marketing costs. The United States Department of Agriculture considered the trend toward cooperation in marketing and distribution as the "most distinct and significant movement in American agriculture in this [postwar] decade."[9]

The postwar years witnessed the most ambitious cooperative venture yet attempted by American wheat growers. They hoped to create in the United States Grain Growers' Company a grain-marketing organization comparable to the United Grain Growers of western Canada.[10] With the support of the Farm Bureau, the Farmers' Union, the Society of Equity, the state Farmers' Grain Dealers' Associations, and the Department of Agriculture, the company was launched in April 1921. By March 1922 more than fifty thousand American farmers had joined. Four months later the company was on the rocks, an audit of its finances revealing a net deficit of $285,000.[11] Efforts at reorganization failed, as a lack of harmony among leaders and excessive organization costs embarrassed the company. The ambitious program was a fiasco, but other and more successful efforts followed on the heels of this disaster.

Canadian farmers expanded their wheat-marketing cooperatives into the famous wheat pools, but only after unsuccessful efforts to continue wartime governmental marketing controls. The fight to retain the wheat board dominated the struggle for economic reform during the immediate postwar years. The farmers were not united in their demands for fixed prices, however, and spoke with confused voices. The Manitoba grain growers and the U.F.A. passed resolutions in their annual conventions of 1919 opposing fixed prices.[12] This official action was extremely unpopular with the rank and file, who believed that adequate prices could be retained only through government intervention. The official pronouncement of the two societies was severely criticized by farmers who strongly

[9] United States Department of Agriculture *Yearbook*, 1925, 17.

[10] The Grain Growers' Grain Company and the Alberta Cooperative Elevator Company joined forces in late 1916 to form this powerful combination. The Saskatchewan Cooperative Elevator Company remained aloof from the merger.

[11] *Senate Document* No. 95, 70th Congress, 1st Session, "Cooperative Marketing," 69.

[12] *Grain Growers' Guide*, 12:6 (February 5, 1919).

suspected that the vote had been "manipulated" by their leaders and did not represent the real wishes of the farmers themselves.[13] The Saskatchewan grain growers reversed the position of the other organizations, and demanded retention of the wheat board.

The Canadian Council of Agriculture added to the confusion by approving a wheat board as a temporary measure, while the principal countries importing Canadian wheat retained government controls over imports and distribution.[14] It affirmed faith in the normal marketing processes and called for their resumption with a return of normal conditions. Again in July 1920 the council reaffirmed its faith in "voluntary cooperative marketing under the control of the participating producers."[15] By October, however, the council had returned to its old demand for the reappointment of a wheat board. Since it was clear that Ottawa would reject this demand, the council appointed a wheat markets committee to explore the possibilities of a voluntary pooling scheme.[16]

The struggle over the wheat board is a story that has been told often. The effects of the agitation on the farmers' movements in western Canada deserve analysis here, however, as a primary cause for political revolt. The refusal of the government to re-establish the board for the 1920 crop was a keen disappointment to western farmers, and convinced them that nothing could be expected from Mr. Meighen's National-Liberal and Conservative party.

The wheat board agitation further revealed a fundamental change in agrarian thinking in the West. Under the impact of the war, western farmers abandoned their frontier and Jeffersonian *laissez-faire* in favor of government intervention in the marketing process. This transition placed the agrarians in an embarrassing position, for it left them advocating free trade on the one hand and favoring government support for agricultural prices on the other. They were dangerously near the position of denying to industry what they demanded for themselves.

Some were aware of this apparent inconsistency and opposed

[13] House of Commons *Debates*, Session 1919, 4:3528; R. C. Henders admitted he had forced reconsideration of the issue in the Manitoba society. *Ibid.*, 3647.

[14] *Canadian Annual Review of Public Affairs*, 1920, 104.

[15] *Grain Growers' Guide*, 13:3 (July 21, 1920).

[16] Patton, *Grain Growers' Cooperation in Western Canada*, 200, 201.

the wheat board as a threat to the moral position of the grain growers. They refused to trade a permanent plea for temporary advantage. One writer in the *Guide*, for example, feared that the agitation for government price fixing revealed that the farmers had advocated free trade from purely selfish motives. "We have regarded ourselves superior to greed . . . but until now the sincerity of the farmers has never been put to test."[17] The spectacle of the farmers demanding protection for themselves and denying it to others was a matter of grave concern. Most western farmers were convinced, however, that theoretical niceties were irrelevant in view of the collapse facing Canadian agriculture. The well-timed report of the Canadian Council of Agriculture's wheat pool committee, which was issued just before the federal election of 1921, revealed that agrarian leaders believed voluntary contract pools to be impracticable. Government action was the only alternative, and the election of a sympathetic government in Ottawa was imperative.

Canadian farmers enthusiastically resumed the crusade against protection and were vigorously opposed by their old enemy, the Canadian Manufacturers' Association. A new enemy, the Canadian Reconstruction Association, had also entered the lists against them in the struggle to sway public opinion. American farmers, on the other hand, were now high tariff in sentiment and pursued an opposite course. Wheat growers in the American Northwest feared their markets would be inundated with a flood of Canadian wheat and sought tariff legislation to protect themselves. The importation of twenty-five million bushels of Canadian wheat into the Minneapolis mills in 1920 was widely publicized and helped win support for the Fordney tariff proposals. The passage of this Emergency Tariff Act in 1921 imposed a duty of thirty-five cents a bushel on wheat coming into the republic. This marked the beginning of a protection policy which sharply restricted imports of Canadian wheat and by 1924 virtually prohibited them.

The adoption of protection by the American Congress worked a hardship on Canadian producers. It laid to rest the ghost of reciprocity, which as late as April 1921 had walked the halls of

[17] Letter of F. Howell, Boissevain, Manitoba, in *Grain Growers' Guide*, 12:34 (March 5, 1919).

Parliament in the form of a proposal by W. S. Fielding to approve the aging 1911 reciprocal trade offer. The adoption of the Fordney tariff also killed an infant movement, for it struck a death blow to a plan looking to cooperation between the farmers' organizations of the two countries.

These efforts reached a highwater mark in May 1920 at the Chicago convention of the National Board of Farm Organizations. Canadian representatives of organized agriculture were enthusiastically received, and a resolution was endorsed recommending "the establishment of an International Committee representing the organized farmers of the United States and of Canada."[18] The convention discussed the economic relations between Canadian and American farmers, and a wide range of subjects of mutual interest to North American farmers was placed before the group. The agenda included discussion of naval disarmament, relations with organized labor, and a proposal to establish an international board of agriculture representing nearly two million farmers to examine all issues affecting agriculture in the two countries.[19]

Though the Chicago conference failed to create a continental movement, it deserves to be remembered. The representatives of organized agriculture recognized deep mutual interests, if only for a passing moment. They discussed economic and political problems with a frankness and a candor that augered well for future relations between the two nations. Canadian delegates, unlike those of the Grange and Patrons of Industry in earlier days, joined the conference with no sense of inferiority or fear of dictation by the larger American societies. In striking contrast to the earlier international conferences, the Canadian farmers were more than "fraternal delegates" who delivered unctuous greetings to fellow farmers in the republic. These men made practical proposals to improve the economic position of the North American farmer.

Canadian representatives were enthusiastic over the progress made in Chicago. R. W. E. Burnaby, one of the delegates, publicly announced, "The purposes of the farmers of both countries are identical, and as soon as we get the cooperation of the new minister to the United States, we will call a special meeting and formu-

[18] *Canadian Annual Review of Public Affairs,* 1920, 120.
[19] *Chicago Daily Tribune,* May 12, 1920.

late plans to combine our organizations."[20] Other Canadian farmers also saw evidences that international agrarian cooperation was in the offing. Charles Harris, a director of the U.F.A., believed that "the time is ripe for us to push our plans for a big international organization of farmers for the purpose of making uniform prices in farm produce."[21]

These were idle dreams. American farmers were now converts to economic nationalism and were the willing partners in the marriage to Republican protectionism. When the Canadian Council of Agriculture met in December, its plans for international cooperation were buried, and in their stead there was talk of retaliation against the American high tariff.

The uneasy wartime truce on the tariff issue in Canada was ruptured by the return of peace. The presentation of the mildly protectionist budget in 1919 provoked the resignation of Crerar from the cabinet and led to the formation of a parliamentary Progressive party of ten members.[22] Many farmers were convinced, however, that the war emergency was not yet over, and the union government's mandate was still in force. Sincere men were divided on this issue, as the defection of R. C. Henders from the farmers' movement clearly revealed. Henders, who for many years had led the Manitoba grain growers and had been elected to Parliament from Macdonald in 1917, believed it his duty to support the union government. In his view 1919 was still a war year and the tariff truce should be continued.[23]

This interpretation was completely unsatisfactory to leaders of the western farmers, who were convinced that attention and flattery from government forces had turned his head and led him into betraying his former colleagues. Whatever his motives, his fate was certain. The grain growers acrimoniously turned on their onetime leader and forced his retirement from the movement. The lines of battle were now clearly drawn. Two decades of anti-tariff agitation had at last created a new parliamentary party whose popular strength was not accurately measured by its handful of representatives in Ottawa.

[20] *Manitoba Free Press,* May 13, 1920.
[21] *Grain Growers' Guide,* 13:31 (January 7, 1920).
[22] House of Commons *Debates,* Session 1919, 4:3193; *ibid.,* 3678.
[23] *Ibid.,* 1:455, 4:3647–51.

The new party inherited a ready-made platform, for the Canadian Council of Agriculture had revised the Farmers' Platform in November 1918. Postwar issues were included, and the new platform was presented to the public as the New National Policy. The inspiration was obvious, and many farmers shared the council's conviction that the old National Policy had outlived its usefulness. Fiscal policy was again the heart of the program, and the tariff remained the symbol of revolt. New issues, born of the war, figured prominently in the new document. New National Policy called for a League of Nations "to give permanence to the world's peace by removing old causes of conflict" and recognized the growth of Canadian nationalism during the war by opposing attempts "to centralize imperial control."[24] The empire was conceived of as a partnership of nations "free and equal."

The problem of the returned war veteran was recognized by the new Farmers' Platform. To the New National Policy, it was the nation's duty to care for her returned soldiers. Demobilization should give preference to length of service, marital status, and industrial needs. Unemployment would be lessened by discharging men only as "it was found possible to secure steady employment." Vocational training, public insurance, and farm lands were proposed to relieve economic tensions accompanying demobilization and to aid in the rehabilitation of the soldier and the nation.

For the first time labor was mentioned in the Farmers' Platform. With an eye to the labor vote, the New National Policy admitted the "very serious problem confronting labor in urban industry." The farmers' solution vaguely called for an attempt by the various units of government to relieve unemployment and hopefully recommended the adoption of "the principle of cooperation as the guiding spirit" in industrial relations.

The crusade for democracy which had been in ferment for twenty years found full expression in the New National Policy. Abolition of titles, reform of the Senate, direct legislation, abolition of patronage, removal of press censorship, declaration of ownership of newspapers and periodicals, proportional representation, prohibition, and the admission of women representatives to

[24] All quotations are from "The New National Policy as adopted by the organized farmers of Canada in the Farmers' Platform" (Winnipeg, 1920).

Parliament—these were demands that had become articles of democratic faith on the Canadian prairies.

The farmers fell like ripe grain before the sweep of the new party. Here was what they had long demanded, and few hesitated to cast their lot with the Progressives. The revolt from the older parties was so complete on the prairies that the *Farmers' Advocate* observed with a good deal of envy, "The political party led by the Hon. T. A. Crerar is the most perfectly organized political machine that exists on this continent. To record a matter of simple fact, it is among the four most perfectly organized political groups that exist in the world today."[25]

The agrarian crusade in western Canada was in the Populist tradition, for it too was in revolt—not only against party controls, but also against the system. The farmers were thoroughly aroused, and few regular politicians on the prairies could hope to escape with their political lives. To the south this form of agrarian political action had passed away, for the new and more powerful farm bloc scheme had replaced it. The postwar decline of the Nonpartisan League, progressivism, and independent farmers' parties deceived many into concluding that the farmers had ceased to act along political lines. Actually, a very strong agrarian radicalism persisted in the Middle Border. Though it now elected Congressmen with party labels, they were committed to an agrarian program. Agrarianism continued its course uninterrupted by the passage of independent parties,[26] and the election of Senators Frazier, Kenyon, Ladd, Capper, LaFollette, Shipstead, and Norris guaranteed that the middlewestern farmers' interests would be attended to in Washington. The organization of the farm bloc in the spring of 1921, with the blessings of the American Farm Bureau Federation, inaugurated a new technique, but the results were much the same for the agricultural community.[27]

Western Canadian agrarians were forced to enter dominion politics as a separate party. The farm bloc idea could not be applied to Canadian politics because the British system of respon-

[25] *Farmers' Advocate*, 56:1265 (November 9, 1921).

[26] Benton H. Wilcox, "A Reconsideration of the Character and Economic Basis of Northwestern Radicalism," unpublished Ph.D. thesis, University of Wisconsin, 1933.

[27] Judson C. Welliver, "The Agricultural Crisis and the 'Bloc,'" *Review of Reviews*, 65:158–65 (February 1922).

sible government could hardly accommodate an agrarian group or bloc whose loyalties extended beyond the party caucus. Postwar political action by North American farmers took diverging courses, though similar fundamental objectives were shared on both sides of the international boundary.

The postwar revolt in Canada first broke out in the East when the United Farmers of Ontario swept into power over the crumbling ruins of the once-powerful Hearst government. Western agrarians were greatly surprised and profoundly encouraged by the "revolution in Ontario."[28] To many a western farmer it seemed incredible that Ontario could lead the way, and that they would now be following a path already blazed by "drowsy, groggy Ontario . . . home of conservatism, sluggishness, slow-thinking, trusts, monopolies, high tariff propaganda . . . soaked in old-partyism, shying at reform."[29]

The U.F.O. victory was easily explained as a result of dissatisfaction with the Hearst government, a protest vote, the prohibition issue, and other purely local questions. Few realized that the rising tide of agrarian revolt was not confined to the prairie provinces, but swept wherever the new commercialized agriculture had developed. The gathering revolt in the West was too easily dismissed as the result of frontier or sectional grievances. The farmers' victory in Ontario revealed that Progressivism was far more than an expression of sectionalism, frontier ignorance, or the discontent raised by imported Populists or Socialists. The farmers themselves were hardly aware of the underlying causes for the widespread revolt. Even today analyses of the agrarian outbreak often ignore the new agriculture and the demands it made for economic controls by governments sympathetic to the farmers' new role in society. The farmers expressed their reactions to these fundamental changes in the striking phrase, "go into politics, or go out of farming."

The victory in Ontario tremendously stimulated the farmers' revolt in the West. The Canadian Council of Agriculture formally recognized the Progressive group in the House of Commons in

[28] *Grain Growers' Guide,* 12:5 (October 29, 1919).

[29] Bert Huffman, "Vagabond Comments," *Western Independent,* 1:2 (October 15, 1919).

January 1920, and the National Progressive party was launched. By March 1920 the new parliamentary group was fully organized in a separate caucus in the House. T. A. Crerar was, of course, its parliamentary leader, and Dr. Michael Clark, the colorful free-trade Liberal from Red Deer, and John Maharg, president of the Saskatchewan Cooperative Elevator Company, were his lieutenants. Though the provincial societies were left with power to determine the exact course of action to be adopted, a clear pattern of progressivism, conforming in spirit to doctrines preached on the prairies for two decades, soon emerged. The new movement was variously denominated as the National Progressive party, the Farmers' party, the party of the New National Policy, the People's party, and the Progressive movement. The last title, perhaps more than any other, defines the true spirit of the revolt. Strict party discipline and party machinery were largely missing, and the revolt, in its larger sense, was not so much a party as it was a movement. A new political movement had thus been born; coming elections were to prove it an unusual infant.

The revolt against "partyism" prevented the adoption of many political practices which experience had shown essential. The emphasis upon local organizations as the source of policy and power kept the organization close to the people, but it also prevented decisive or uniform action. Provincial particularisms weakened the movement and created deep fissures. Divisions in the ranks quickly demonstrated the inability of the movement to discipline its members or present a united front in dominion politics.

These weaknesses were clearly demonstrated during the bitter controversy that raged over the nature of the movement. Was the Progressive movement a class party or not? Political opponents were unanimous in their condemnation of the farmers' program as an effort to set class against class. Here was a ready-made issue for the party stalwarts, and they played it to the full. The farmers were badly divided on the question, and no other issue was as hotly contested as this one. Many frankly admitted they were entering politics as a class; others vehemently denied it. A number of prominent agrarian leaders left the movement in disgust as protagonists of one concept or the other gained the upper hand in local or provincial societies.[30] The United Farmers of Ontario were soon hopelessly split as J. J. Morrison, the venerated farmers' leader, led the attack against Premier Drury's "open door" or "broadening out" policy.

The disruptive controversy was never settled. The farmers often thought and acted with a strong class consciousness, but the position of the Progressive movement was far from that of a narrow class party seeking to exploit other groups for selfish advantage. If their opposition to the tariff was based on class interest, then it may be justly said that those who favored protection were equally guilty. The *Manitoba Free Press* rightly believed that "in its origins this was, of necessity, a class movement, but . . . the movement is rapidly evolving from the class stage to the status of a true political party with a Dominion-wide programme and a national vision."[31] Not all farmers, however, grasped the necessity of compromising their demands to reconcile sectional and class interests in the arena of national politics.

The inability of the movement to formulate a clear-cut labor policy was another indication of the confusion within the ranks. The Progressive movement failed to create any formal alliance with labor, though local cooperation was often the rule. Farmers' journals emphasized the urgency of farmer-labor cooperation. The *Grain Growers' Guide* and the *Western Independent* frequently cited the cooperation of the U.F.O. with organized labor as an example for the West. Farmer cooperation with labor was essential in industrial Ontario to keep the Drury government in power. In the overwhelmingly rural society of the prairie provinces, however, no such imperative forced unity, for the farmers already held a preponderance of political power.

Responsible leaders sought to end the confusion by calling for uniform action. The *Guide* believed that the primary producers would have to unite forces to secure justice and denounced "agitators" who sought to deceive the farmer into believing that a wide gulf of differences separated him from the industrial worker. The *Western Independent*, organ of the U.F.A. political association,

[30] The erratic James Weir, for example, charged the U.F.A. with "Sovietism," and gave aid and comfort to the farmers' enemies with his bitter attacks on H. W. Wood.

[31] *Manitoba Free Press*, December 3, 1920.

strongly advocated cooperation to defeat the common foes of "exploitation and autocracy."

This cooperation was greatly weakened in practice by the belief that a basis for political cooperation existed only after each class had elected its representatives to Parliament.[32] The grain growers' associations were keenly aware of the problem, but seldom passed the stage of appointing committees "to study ways of bringing about a better understanding."

Agrarian sympathy for organized labor was shocked into open hostility by the Winnipeg general strike of 1919. The violence and radicalism of the strike turned many farmers against organized labor, and cooperation on a widespread scale was impossible thereafter. Spokesmen for the grain growers were united in their disapproval of the methods adopted by the strikers. J. B. Musselman damned the strike as a "revolutionary movement for the confiscation of all private property, including farm land" and warned that labor would receive no support from Saskatchewan grain growers.[33] The *Guide* adopted a similar editorial policy. Inflammatory talk, Bolshevism, sympathy strikes, and violence were decried as vicious and undemocratic.[34] Thereafter, the *Guide* seldom called for any alliance with labor. Western farmers generally accepted the dominion minister of labor's definition of the affair: "This is not a strike but a plain, ugly revolution."[35] Most farmers believed the ready explanation that the ringleaders of the strike were foreign agitators, and the Manitoba jury of farmers which condemned three of the labor leaders to prison was applauded throughout the West.

Since neither Progressive doctrine nor political necessity dictated a general alliance in the West, relations with organized labor were highly localized. Cooperation often existed, but it was limited in nature, and usually raised severe opposition among less tolerant farmers who looked upon the laborer's wages as a primary cause for the high cost of living. T. A. Crerar, who was as close to being the spokesman for the party as anyone, frequently

[32] *Western Independent*, 1:6 (October 22, 1919).

[33] *Grain Growers' Guide*, 12:32 (June 25, 1919).

[34] *Ibid.*, 12:6 (May 21, 1919). The *Guide* categorically condemned the strike: "The Winnipeg strike failed completely because it was wrong. Many of its leaders preached the worst doctrines of Bolshevism, confiscation and rule by force."

[35] House of Commons *Debates*, Session 1919, 3:3008.

repudiated the idea that the Progressives were a class group in the sense that labor was, and often advanced the proposal that the new party was in an excellent position to hold the balance "between capital on the one side and labor on the other because the farmers were both capitalists and laborers."[36] The federal election of 1921 revealed the nature of the Progressive-labor relations. During that crucial period, the farmers joined with labor in some constituencies, and fought the workers' party in others.

These fundamental differences of opinion soon divided western farmers as badly as they had the United Farmers of Ontario. This dichotomy of growth on the prairies produced the "Manitoba view" on the one hand and the "Alberta group" on the other, each with its eloquent protagonists. T. A. Crerar personified the ideals of the former; Henry Wise Wood represented the latter. Both were leaders of unusual ability, possessed of unquestioned integrity and moral courage. Each enlisted the devotion and enthusiasm of a large following of prairie farmers. Much of the success of the United Farmers in business as in politics rested, in last analysis, upon the wisdom and integrity of these two men.

Thomas A. Crerar, of course, was a dominion figure by the time the Progressive movement took shape in the West. He had been associated with the grain growers' movement from its earliest years. As head of the United Grain Growers' Company and as minister of agriculture in the union cabinet, he had often demonstrated unique qualities of leadership and sound business judgment. Few men possessed his understanding of the problems of the Northwest wheat belt, for his experience in production, marketing, and government had provided an unusual background.

Henry Wise Wood's rise to leadership, on the other hand, had been more recent. His first office with the U.F.A. was as late as 1914, when he was elected vice-president. The death of President James Speakman the following year elevated Wood to the presidency and lifted him from the obscurity of Carstairs. In 1917 he was elected president of the dominion-wide Canadian Council of Agriculture and later he was a member of the dominion Board of Grain Supervisors. His American background was a serious handicap and exposed him to the bitter attacks of enemies of the farm-

[36] *Manitoba Free Press,* November 18, 1920.

ers' movements who were not above embarrassing him for this accident of birth.[37] They quickly succeeded in labeling him "the man from Missouri," and by that appellation he was popularly known throughout the dominion.

The "Manitoba view" considered the Progressive movement as the "forward looking" or "progressive" wing of liberalism. The Progressive movement, it was hoped, would convert the Liberal party to the principles of the New National policy, while Liberal protectionists would be forced to flee this revitalized liberalism to the more hospitable atmosphere of Conservatism. Many feared, moreover, that a separate farmers' party would divide the low tariff strength, and thus throw political victory into the hands of the protectionist forces. The powerful *Manitoba Free Press* encouraged the Progressive movement as "genuine liberalism," and predicted that it would be the foundation for "a great national party which, whatever its name, will be in its policies liberal in the true acceptance of the often ill-used term."[38]

The National Progressive party, as conceived by this wing of the western farmers' movement, would make its appeal to all classes of Canadian society and seek its strength in the cities as in the country.[39] Crerar foresaw the danger of dividing Canadian parties along class lines and deplored the tendency of making the farmers' movement into a class party selfishly seeking advantage for the farmer at the expense of other elements in Canadian society. This view demanded little more than a revolt from the rule of the party caucus and was opportunistic in its efforts to remove the heavy burden of the protective tariff from Canadian agriculture.

The Alberta "group government" agitation, on the other hand, was one of the few farmers' movements to develop a systematized politico-economic philosophy. There was at the time, and there remains, much misunderstanding of Henry Wise Wood's "economic group government" concept. Many professed to see Bolshevist or Marxist influence in the doctrine. Actually, Wood simply reduced to a philosophic system the physiocratic ideas he had heard since

[37] The *Calgary Daily Herald*, for example, once called Wood "a bad Yankee that has come to destroy British responsible government."
[38] "The National Progressive Party," *Manitoba Free Press*, December 3, 1920.
[39] "The New National Policy," 8.

his youth on a Missouri farm.[40] He, like most North American farmers, had early observed the efficiency of the industrialists' "class organizations." On both sides of the international boundary, farmers envied the powerful influence over government exerted by such organizations as the Canadian Manufacturers' Association and the National Association of Manufacturers. They had witnessed the rapid growth of trade unions into powerful class organizations, and many believed, as a result, that the farmers, alone unorganized, carried the burdens of the economic system.

To these commonplace observations, Wood applied ideas that had matured out of his early training under Professors J. C. Reynolds and B. H. Smith at Christian College in Canton, Missouri.[41] There he had learned that competition was a "false social law" and that no social system based on competition could ever reach perfection. Cooperation was the true social law, and the inspiration for his philosophy was not the Communist saint, Karl Marx, but the Christian saint, the Apostle John.[42] Though Wood was well acquainted with Marxist doctrine and had read everything by Karl Marx he could lay his hands on, he vigorously opposed the Communist philosophy. He was firmly convinced that the Marxist dialectic could not be applied to the agrarian movement, and he wanted no part of that philosophy in his own planning.[43] Cooperation, not class domination, was the ultimate goal, but cooperation could be secured only through the organization of efficient class groups which, when fully organized, placed competition on the highest possible level where it was merged in cooperation. The heart of Wood's doctrine was the proposition that "power treats power on equal terms, in the hope of an equitable adjustment."[44]

Practical experience as a farmer in Missouri confirmed these judgments and strengthened his faith in cooperation. In an early

[40] "Wood Papers," manuscripts in the possession of Leonard D. Nesbitt, Calgary, Alberta. Wood once declared, "I have always had the idea of an organization among farmers ever since I was a member of a little Missouri local." He referred to the Farmers' and Laborers' Union in Ralls County, Missouri.

[41] Now Culver-Stockton College, Canton, Missouri.

[42] Wood was a thorough student of the writings of Saint John. At the time of his death he was writing a book setting forth his social philosophy. The title of the proposed volume was "The Social Aspects of the Life of Christ."

[43] Letter to the author from John S. Wood, St. Louis, Missouri, February 17, 1947.

[44] H. W. Wood, "The Efficient Citizenship Group," *Grain Growers' Guide*, March 22, 1922.

brush with the powerful Bell Telephone System, Wood demonstrated his keen sense of social justice and his determination to challenge the economic despotism of monopoly. The refusal of the Bell system to give farmer-subscribers adequate service at reasonable rates moved him to organize a farmers' mutual telephone company. The farmers' company was so successful that the Bell system was forced to give the farmers the service and the rates they believed reasonable.[45] This experience revealed to Wood the power of organization, and the potential effectiveness of economic organization by the farmers. His companions in the telephone system were witnessing the schooling of the young Missourian who was destined in later years and on distant prairies to lead in organizing extensive cooperative activities and eventually to head the powerful Alberta Wheat Pool in western Canada.

Wood traced the historical background for the emergence of the cooperative spirit from the earliest days of the race, when the "law of ruthless competition" placed men on an animal level of existence. Gradually, however, the law of cooperation had replaced the law of competition as the weak were forced to form groups for self-preservation.[46] The efficiency and not the size of the group protected the individual from his predatory neighbors. The only basis for an efficient group organization in modern society was economic, for any other motivation lacked the stability and strength provided by a community of economic interest.[47] Ideas were often a "froth" that intrigued men, but economic interests were the forces that impelled groups to action.[48] This struggle, however, was basically ethical; it was one in which the forces of evil sought to protect their privileged position against the growing democratic power of the common people. "This showdown," wrote Wood, "between the Mammonistic and the humanistic forces will not develop to the final stage until the elements of each force are thoroughly mobilized through systematic organization. The Mam-

[45] Letter to author from Judge W. T. Ragland, Orange, California, January 14, 1947. Judge Ragland, a former justice of the Missouri Supreme Court, was one of Wood's closest friends.

[46] There is no evidence that Wood was familiar with the work of Kropotkin despite the similarity of ideas.

[47] H. W. Wood, "The Significance of Democratic Group Organization," pt. IV, *U.F.A.*, 1:5 (April 15, 1922).

[48] William Irvine, *The Farmer in Politics*, 163.

monistic forces have already made much progress in the process of mobilization."[49]

Wood was critical of the plutocracy and denounced the inequalities in Canadian and American society which his sensitive spirit so often felt. With the evangelistic power of a "Sockless" Jerry Simpson, a Mary Elizabeth Lease, or a William Aberhart, whom he resembled in this respect, he defined the inarticulate feelings of thousands of prairie farmers and stirred them to action with eloquent sincerity. The role of prophet and philosopher was his; to more practical men he wisely left the administrative details.

The collapse of the Populist crusade had made a deep impression on Wood, leaving him forever soured on the idea of a farmers' political party. He had suffered bitter disappointment in the betrayal of the farmers' cause by his hero, U. S. Hall, the able but ambitious congressman from Missouri. Flattery and political preferment had turned the head of this one-time champion of Missouri farmers and transformed him from a crusading liberal to a reactionary. Wood's deep suspicion of individuals in the farmers' movements who cherished political ambition was the heritage of this early experience.[50] Critics who feared Wood as a political demagogue or as a transplanted Populist ignored his "unalterable opposition to the creation of a farmers' political party,"[51] and overlooked the fact that he was morally incapable of demagoguery.

The Alberta leader never endorsed the Progressive party with any enthusiasm. He was, in fact, deeply critical of the agrarian tendency to pursue "that phantom of creating a new political party" which had wrecked so many farmers' movements. He undoubtedly feared that the Progressive party placed in jeopardy the cooperative and economic activities of the United Farmers movements. This strong anti-third-party sentiment was similar to that which had been accepted by many American farmers as a result of the collapse of Populism in the 1890's and the Progressive fiasco in 1912. Wood believed that there was no occasion for third parties if farmers would "go to their party caucus and help send dele-

[49] H. W. Wood, "In Defense of Group Politics," *Canadian Forum*, 3:72 (December 1922).

[50] Memorandum to the author from John S. Wood, St. Louis, Missouri, February 18, 1947.

[51] *Grain Growers' Guide*, 14:27 (October 5, 1921).

gates to their party convention who would nominate the right candidates."[52] The relentless pressure of rank and file opinion forced the organization into politics, overruling the judgment of Wood. He insisted that the U.F.A. had entered politics as an economic organization, not as a political party. Years later, he admitted that he had used his influence to prevent the venture into politics and testified, "I kept out as long as I could."[53]

Henry Wise Wood was an idealist and a perfectionist. By conviction and by family tradition he was a Jeffersonian liberal who believed in the ultimate perfectibility of society, though improvement and perfection were the work of centuries. When younger colleagues grew impatient with the slow progress of the farmers' movement, he warned them not to take "too short a view in the arc of time."[54] He believed that the ultimate victory would rest with democracy because of the moral nature of the universe. "The Triumphant Democracy cannot fail," he once declared, "because the Supreme Power . . . has this work in hand and will not let it fail."[55]

The practical application of "group government" to Canadian politics was explained to western farmers by the facile pen of William Irvine, who, sensing the coming farmers' revolt, was now to be numbered among Wood's disciples. In 1920 he repudiated his stand of the previous year that no farmers' party had a monopoly of democratic principles or should refuse membership to urban citizens. He apparently forgot his own warning that it was unlikely that any strictly "farmers' party will be any more in demand than Tory or Grit."[56] In defense of Wood's doctrines, he published *The Farmers in Politics,* in which he insisted that the U.F.A. had developed as the inevitable result of fixed economic and social laws. The farmers had been forced into class organizations, he contended, by the "social law" of self-preservation, but only they "of all the economic groups of Canada, have discovered the higher law of cooperation . . . between competing groups."[57] The group was

[52] *Ibid.,* 10:12, 13 (June 20, 1917).

[53] "Wood Papers," manuscripts in the possession of Leonard D. Nesbitt.

[54] *Ibid.*

[55] H. W. Wood, "The Efficient Citizenship Group," *Grain Growers' Guide,* March 22, 1922.

[56] *Alberta Non-Partisan,* 3:6 (June 5, 1919).

[57] Irvine, *The Farmers in Politics,* 147.

not the end but the beginning of democracy, for cooperation had first to be learned in the group.

Irvine saw the practical problem that faced "group government" while seeking power in the British two-party system of responsible government. He had wrestled with the same problem in the Nonpartisan League, where he had worked out a solution. He explained the U.F.A. position by pointing out that group government implied that each group would be represented in the cabinet. What the U.F.A. really meant was "group representation" similar to the "corporate state," though without the elements of coercion and dictatorship which Mussolini applied to it in Italy.

U.F.A. leaders were convinced that "group representation" would not impair the British principle of responsible government. They contended, moreover, that "an artificially fixed opposition" was a fundamental weakness in the British system, based as it was on the implication that everything attempted by the government was wrong.[58] The likelihood of "cabinet domination" over the parliament would be reduced if no government could be defeated "except by a direct vote of want of confidence."[59] Government by group representation, they believed, would create a flexible government, and would prevent any one class from dominating Canadian politics.[60] Farmers would then get the justice denied them under the existing government.

The Alberta view was that of the "true spirit of Progressivism," which eschewed political parties and revolted against the moral bankruptcy of "partyism." Doctrinaire and heterodox as it was, the movement produced an effectively disciplined organization which dominated Alberta politics from 1921 until the headier wine of Social Credit intoxicated Alberta farmers with its promises in 1935.

The scattered federal by-elections of 1920 and 1921 were harbingers of the political revolt to come. The election of three Progressives in eastern constituencies confirmed the politicians' fears that the farmers' party was dangerously strong. Liberals sought to head off the stampede of farmers' votes by pleading that the two

[58] *Ibid.*, 239; House of Commons *Debates*, Session 1922, 1:213–17.
[59] United Farmers of Alberta, *Annual Report and Yearbook*, 1921, 40.
[60] Irvine, *The Farmers in Politics*, 207.

parties "hold similar beliefs and have similar objects." [61] Prominent Liberals stated the belief that "the Farmers' party and the Liberal party are one and the same," and they were really "two names for the same thing." Mr. W. L. Mackenzie King, now head of the dominion Liberal party, sought to avoid a direct clash with the farmers, and Liberals opposed farmer candidates with manifest reluctance. During the Stormont-Glengarry by-election, for example, Mackenzie King made it clear that he hesitated "to place the Liberal party in a position antagonistic to the farmers." [62] Hope lingered in Liberal breasts that the farmers' revolt could be diverted, and the farmers lured back to their old Liberal loyalties.

Progressive strength on the prairies was dramatically revealed in the smashing victory of O. R. Gould, the farmers' candidate in Assiniboia, over W. R. Motherwell, the prominent Liberal and early leader in the grain growers' movement. The Saskatchewan farmers rejected Motherwell so completely that he lost his election deposit. The election of Alex Moore in the provincial riding of Cochrane in Alberta was equally convincing evidence that the U.F.A. was also a power to be feared. These victories cheered Progressives and encouraged them to hope that the next federal elections would leave them in control in Ottawa. Progressive candidates rolled up impressive victories, for the farmers' revolt was now in full tide and it threatened to sweep away the old party landmarks in the West.

The farmers' revolt reached its climax in the federal election of 1921. The campaign was long, tedious, and bitterly fought on issues that had troubled dominion politics since 1918. Prime Minister Meighen courageously defended protection, and lashed out against the "free-wreckers" in the "Wood-Crerar" party whose loyalty to Canada he challenged and whose ideas, he announced, were foreign-inspired. The National-Liberal and Conservative party hoped to win with time-tested appeals to Canadian nationalism and anti-Americanism. Canadians were warned that the danger was as great as 1911. Huge posters, picturing the Canadian beaver secure in his protection hut with the fierce and menacing

[61] Speech of W. L. Mackenzie King, October 8, 1920, at Camrose, Alberta, *Canadian Annual Review of Public Affairs*, 1920, 433.

[62] *Grain Growers' Guide*, 12:5 (September 3, 1919).

American eagle ready to seize the hapless animal should he emerge, were erected throughout the dominion. Conservatives were wasting their breath this time. The tide was running out, and the old clichés could not check it.

Mackenzie King's Liberal party stood squarely on the platform of 1919, which like so many political platforms meant all things to all men. Liberals labored to keep the farmers loyal and posed as their faithful friends. Both Liberals and Progressives profited by the Riordon tax scandal revelations, which convinced many Canadians that the Meighen government was inefficient, if not actually dishonest.

Progressives entered the election confident of victory. They based their campaign on the New National Policy which by this time was widely known throughout Canada. The Progressives waged a strange battle. They signed recall statements, they publicized their party contributions, and they committed nearly all the mistakes that professional politicians avoid. There was no uniform pattern in their campaigning. In most constituencies, Progressive candidates denied class partisanship and repudiated the Wood doctrines. In others, however, candidates frankly admitted they stood for class representation. Relationships with organized labor were even more confused. In most constituencies there was no alliance with labor since none was needed to insure victory. In Alberta, labor and the U.F.A. candidates often fought bitter fights, while in adjoining constituencies the two groups united forces.

Curious situations often developed. In Strathcona, for example, Rice Sheppard, though a U.F.A. stalwart, campaigned as a labor candidate and attacked Daniel Webster Warner, the U.F.A. candidate, with unusual venom and bitterness. In Southern Alberta, M. F. Finn, the labor candidate, and L. H. Jeliff of the U.F.A. traded blows in Lethbridge, while in the adjoining constituencies of Calgary East and Calgary West, organized labor and the U.F.A. united to endorse William Irvine and J. T. Shaw.[63]

Progressive candidates, with a few exceptions, were not impressive. Most of them were politically inexperienced, rather colorless, and without any great ability. Of those who had already been in Ottawa, several had revealed shocking incompetence. Some even

[63] *Manitoba Free Press*, December 1, 1921.

had difficulty in securing renomination from their farmer constituents. O. R. Gould, whose dramatic victory in the earlier election had so stirred the West, faced a lukewarm convention, and salvaged his renomination only after a fighting, last-minute speech.[64] Western farmers concentrated on the vindication of principles in the campaign and paid slight heed to the quality of their candidates. The movement later reaped a bitter harvest for this tragic oversight.

The election results revealed the magnitude of the agrarian revolt in western Canada. The Progressives swept the prairies with a clean broom as regular party politicians ignominiously forfeited their deposits in one-sided contests. Not a single government candidate survived in the West, though R. B. Bennett waged a desperate fight in Calgary West and lost by only 16 of the 16,181 votes cast.[65] Of the forty-three prairie members of Parliament, only four were not Progressives, and of these two were independent candidates. The Progressive revolt joined a resurgent Liberalism to smash the Meighen government in the most humiliating defeat yet suffered by a dominion cabinet. The Prime Minister and seven members of his cabinet were among the casualties as the Progressives captured sixty-five seats in the House of Commons.

The party of the New National Policy emerged triumphant from the election as the second largest group in the House. The revolt had swept the farmers into positions of power and responsibility. How they would treat this new-found power, and whether they would measure up to their responsibility, only the months and years ahead could reveal. One thing was certain. The seeds of discontent planted at the turn of the century in those hot-beds of prairie democracy, the grain growers' locals, had matured twenty years later into a political revolt that was fated to influence the course of Canadian politics.

[64] *Grain Growers' Guide*, 14:4 (October 26, 1921).
[65] *Sessional Papers*, 1922, no. 13, 479.

IX

The Farmers in Politics

The agrarian revolt had worked its will and the farmers' party vaulted into power over the wreckage of the older parties in the West. The party of the New National Policy had won striking victories and its future seemed altogether promising. Hopes were widespread on the prairies that a new era in Canadian politics had been born.

Echoes from the election had scarcely died out when disquieting rumors swept Canada that Mackenzie King and certain Progressive leaders were engaged in "negotiations" looking to a coalition of Liberal and Progressive forces. The political analyst for the *Canadian Forum* expressed a common belief when he wrote that "the *grande passion* of the Premier for a Progressive marriage is a central factor in our politics."[1] Certainly, the prime minister seemed anxious during the months following the election to bring certain sympathetic Progressives into the Liberal cabinet. Mackenzie King looked upon the majority of Progressives as an "advanced liberal group," and hoped to entice two or three of their representatives into his cabinet. To this end, conversations were held with prominent Progressives during 1922. The prime minister publicly denied, however, any desire for a "coalition" or "compromise" cabinet,[2] and was apparently unwilling to grant those concessions that Progressives believed essential to the preservation of the identity of their movement.

The proposed coalition quickly became a burning issue with Progressives. The course of the negotiations in Winnipeg, Toronto, and Ottawa, the extent of promises given or understood, and the commitments of party leaders were unknown to the rank and file,

[1] *Canadian Forum,* 3:37 (November 1922).

[2] T. A. Crerar denied this in the House. "When he [Mackenzie King] states that he made it quite clear in any proposals or suggestions he made, that this ministry was only to be a Liberal ministry, I must say that I did not so understand it." House of Commons *Debates,* Session 1922, 1:51.

and remain obscure. Western Progressive members of Parliament immediately convened in Saskatoon to discuss the Liberal offers. After vigorous and animated debates, they decided that their leaders were at liberty to go into a coalition "on their own responsibility."[3] Their rather negative stand was modified, however, by a firm resolve that the movement must retain its identity. This decision was inevitable. The revolt had swept the West as a protest against the iniquities of both the traditional parties, and Progressive doctrine could hardly be reconciled with a union with either of the "corrupt" and "undemocratic" parties. The new members of Parliament were not convinced, moreover, that entry into the King cabinet would force the government into Progressive channels. They looked askance at powerful protectionist Liberals in the cabinet, and particularly feared the influence of Sir Lomer Gouin, whom they regarded as the high priest of Liberal protectionism. The Progressive leaders were badly divided by the coalition issue, and the opening wedge of disunity in Progressive ranks was revealed.

The popular reaction within the Progressive movement was bitter opposition to any capitulation to the Liberals. It was "unthinkable to the rank and file of the S.G.G.A. that Crerar could even negotiate," declared their aggressive secretary, J. B. Musselman.[4] The *Grain Growers' Guide,* though a Crerar partisan, believed "an independence of organization" must be retained, but did criticize those farmers who questioned the good faith of the prime minister's offers.[5] Henry Wise Wood and the U.F.A. were uncompromisingly opposed to fusion with the Liberals on any terms. "No elected member . . . is entitled to commit himself to so complete a reversal of the plan of democratic action he was elected to maintain," wrote W. Norman Smith, the able editor of the *U.F.A.,* the new organ of the Alberta society.[6] President Wood explained that coalition was impossible and cited as a grave warning the fate of American agrarian movements which had merged with established parties. The U.F.A. president believed that the American example, "more than anything else," influenced the course taken in preserv-

[3] *Ibid.,* 36.
[4] *Grain Growers' Guide,* 15:8 (January 4, 1922).
[5] *Ibid.,* 15:5 (January 4, 1922).
[6] *U.F.A.,* 1:17 (November 1, 1922).

ing the identity of the farmers' political organizations in western Canada.[7]

Rumors circulated freely on the prairies that coalition was imminent. Many farmers were convinced that some of their leaders were prepared to desert them for positions in the Liberal cabinet. Robert Gardiner fed the flames of suspicion when he announced to the U.F.A. convention in January that "there was no question in my mind but that Mr. Crerar was there [at the Saskatoon conference] for the purpose of forming a coalition of the Progressives with the Liberal party."[8]

Prominent Liberals and Progressive ex-Liberals were reluctant to admit defeat in their efforts to reach an understanding, and further negotiations were conducted late in 1922 through A. B. Hudson, the Winnipeg independent Liberal. Opposition to fusion continued to grow in Progressive ranks, and all negotiations were doomed to failure. When the *Grain Growers' Guide* excoriated Prime Minister King as a "timorous mediocrity," the eighty thousand readers of this authoritative organ could hardly look with favor upon the Liberal leader. Still, hope lingered that eventual union between moderate Progressives and low-tariff Liberals could be effected. The *Manitoba Free Press* expressed these hopes when it announced that "a fusion between Progressives as a whole and the Liberal party as a whole is quite beyond the range of possibilities. But a union between the members of the Liberal Parliamentary party who happen to be Liberals and Progressives, less an extremist fringe, for a definite purpose, is conceivable."[9] This was a restatement of "Manitoba Progressivism," which always cherished the hope of capturing the Liberal organization in the interests of "genuine Liberalism."

The failure to accept the Liberal offer may have removed the golden opportunity to shape the destiny of the Liberal party, as many ex-Liberals in the Progressive party professed to believe. It is more probable, however, that the Progressives judged the political realities accurately and adopted the wiser course. Sir Clifford

[7] H. W. Wood, "The Canadian Council of Agriculture and Political Action," *U.F.A.*, 2:9 (May 15, 1923).

[8] House of Commons *Debates*, Session 1922, 1:36.

[9] *Manitoba Free Press*, August 5, 1922.

Sifton, whose intimate knowledge of the politics of the period produced many sound political judgments, shared their doubts that "any good would come from the absorption of one or two farm leaders," though he did favor the coalition if it could be formed.[10] The controversy widened the breach between the Wood and Crerar factions, and drove a dangerous wedge of suspicion between leaders and the rank and file, who were fearful that they might be betrayed. The continued rumors of fusion greatly weakened the moral position of the party of protest.

Progressive leaders were sailing an uncharted course as a third party. The decision to reject their earned position as the official opposition in the House of Commons headed the farmers' party toward the rocks of political frustration and impotence where it eventually went to wreck. In this, the Progressives shared the tendency to evade political responsibility that was characteristic of previous farmers' parties. Wedded to the balance of power concept as they were, Progressives could think only in terms of supporting their legislative principles as initiated by the older parties. Mr. Crerar made this clear in his initial parliamentary address as leader of the triumphant party. "We are here," he declared, "for the purpose of furthering certain principles . . . we are prepared to assist the Government in giving the country the policy we think we should have; but we are prepared to oppose the government when we think it is not giving the country that policy."[11]

This fateful decision fitted the doctrinaire position of western Progressives who were in revolt against the evils of partyism, but it was ill-adapted to the realities of political life. The refusal to become a parliamentary party seeking office robbed the Progressives of any degree of permanence in the Canadian political structure. To the average voter, this decision seemed to deny to the victor the exercise of power that had been the object of contest. Why should he waste his vote supporting a party which would reap none of the fruits of victory?

Political disintegration naturally followed the self-denying decision to avoid political responsibility. Many who hoped to see the

[10] John W. Dafoe, *Clifford Sifton in Relation to His Times,* 485.
[11] House of Commons *Debates,* Session 1922, 1:52.

Progressive opposition in the House force the Liberals and Conservatives into an uneasy alliance were keenly disappointed.[12] They feared that their party had destroyed its opportunity for placing the radical economic doctrines of the New National Policy into the political foreground, thus driving a wedge between genuine progressivism, irrespective of party lines, and real conservatism.

Though the failure to become the official opposition in the House of Commons was a tactical error of tragic consequence, the rank and file did not see it at the time. For many years Progressives had thought in terms of wielding a "balance of power" in Parliament; by the time victory distinguished their movement this "fixed idea" seriously limited their parliamentary horizon. The balance of power concept was an illusion, for it depended upon a well-disciplined Progressive group and it could easily be turned against them by the Liberals in power. Except for isolated periods of parliamentary crisis when the life of the Liberal government was threatened by a declining strength, the Progressive representation in the House exercised little real political power. The result was a harvest of suspicion and disrespect from the older parties which accentuated the deep-seated inferiority feeling of many Progressive legislators. They were not playing the game as the politicians and the public expected it to be played, and they paid a high price for departing from established parliamentary tradition.

Parliamentary experience soon taught Progressives that the balance of power doctrine had lured the movement into a false sense of security. T. A. Crerar warned the U.F.M. convention in 1923 that the balance of power strategy was unsatisfactory and had not worked in the interests of Canada.[13] Robert Forke, then leader of the Progressive faction in the House, complained, "Some people imagine that to hold the balance of power is a good position to occupy in this House. I am not one of those. I do not want to . . . simply sit and wait . . . I should like the privilege and oppor-

[12] "If Mr. Crerar had not feared his fate too much and refused after the 1921 election to reject for his party the role of official opposition, by this time the two ancient parties would have been welded together." *Canadian Forum*, 3:40 (November 1923). See also J. A. Stevenson, "The Political Situation in Canada," *Contemporary Review*, 129:455 (April 1926).

[13] *Grain Growers' Guide*, 16:20 (January 17, 1923).

tunity of having at least some share in formulating the principles and ideas that shall come before this House."[14]

The ambition to create a third party unfettered by relations with the older organizations was further handicapped by the large numbers of Progressives who retained a sneaking regard for their old party. Most of the Progressive leaders in Manitoba and Saskatchewan were ex-Liberals. The balance of power position was weakened by the knowledge that many Progressives were temperamentally unable to use the bargaining power of their position when it meant close cooperation with their old enemies, the Conservatives. This parliamentary attitude gave rise to the criticism that the Progressives were nothing more than advanced agents of liberalism, or as "Gloomy Arthur" Meighen strikingly phrased it, the "dilapidated annex" or "foreign legion" of the Liberal party. T. A. Crerar's open friendship with liberalism was bitterly criticized by doctrinaire Progressives, and Robert Forke was widely known as "the Liberal turned Farmer."

Other critics cited the role of the *Manitoba Free Press* as further evidence of the close relationship with the Liberal party. This influential organ of western liberalism displayed a sympathetic attitude toward progressivism, and John W. Dafoe, its brilliant editor, was father confessor to many Progressive leaders. "The *Manitoba Free Press*," wrote the editor of the *Farmers' Advocate*, "is telling Mr. Crerar just what to do. And Mr. Crerar listens to the *Free Press*."[15] The strong feeling of ex-Liberals for their old party convinced many that the Progressives could not become "a vigorous, independent party of protest and constructive radicalism."[16] The growing conviction that the Progressive movement was no longer a fit vessel for the cause of reform was a fundamental reason for the deep dissatisfaction of the radical wing which soon left the Progressive household to seek its political fortune as the "Ginger Group."

The Progressive parliamentary party was critically weakened through the years by a lack of discipline and by an ineffective leadership. The nature of the movement made effective party dis-

[14] House of Commons *Debates*, Session 1923, 1:231.

[15] *Farmers' Advocate*, 56:1473 (December 21, 1921).

[16] J. A. Stevenson, "Strains on Canadian Confederation," *New Republic*, 41:166–68 (January 7, 1925).

cipline impossible, for the Progressive revolt placed the emphasis on the control of the legislator by his constituents rather than by the party caucus. This was clearly demonstrated as early as November 1922, when the Progressives gathered in Winnipeg to plan their national program. A resolution calling for the formation of a federal coordinating agency touched off the clash of rival philosophies. The U.F.A. delegation of Robert Gardiner, Joseph T. Shaw, Henry E. Spencer, Alfred Speakman, W. T. Lucas, D. M. Kennedy, George G. Coote, and L. H. Jeliff threatened to bolt if such action were taken. They forced a compromise resolution which provided that the provincial organizations would hold conferences and then lay their decisions before the constituencies.[17] This dissension continued to grow until eventually it ruptured the party and reduced the Progressive movement to a purely sectional protest.

The Progressive group harbored a movement within a movement, for the U.F.A. members of Parliament never considered themselves Progressives. They were called Progressives when they reached Ottawa, but they had not campaigned in Alberta under that banner.[18] They looked upon the National Progressive party as just another party possessing all the evils of the system against which they were in revolt.[19] The Alberta society opposed the formation of a national party and considered such a move as the determined effort "by certain politicians to destroy the democratic bases" of the farmers' movements.[20] The U.F.A. made it clear that if Progressives organized as a political party they would represent "no one but themselves." H. W. Wood warned his colleagues that the drift toward party politics was undemocratic and violated every principle of group organization.[21] Thus the Progressive movement entered the halls of Parliament a divided party, and the divisions multiplied as the years passed.

The Canadian Council of Agriculture might have imposed a degree of unity upon the Progressive movement, but it was now dominated by H. W. Wood and the doctrinaire wing of the move-

[17] *Grain Growers' Guide*, 15:5 (November 15, 1922).

[18] "The term 'Progressive' has been used loosely by many newspapers to indicate U.F.A. candidates, but it is of course incorrect." *U.F.A.*, 4:3 (October 17, 1925).

[19] *Edmonton Journal*, July 14, 1922.

[20] *U.F.A.*, 1:3 (July 15, 1922).

[21] *Ibid.*, 1:3 (August 15, 1922); *ibid.*, 1:1 (September 1, 1922).

ment. Moreover, there was a growing resentment because the council, which represented commercial companies as well as the United Farmers organizations, was playing an active role in politics.[22] The council, therefore, rather than furnishing decisive leadership, added to the confusion by announcing that it would no longer "take part in the conduct of any campaign." Political decisions were to be left in the hands of the provincial organizations.[23] Thus the one organization to which many looked for political guidance to lead the Progressive movement out of the deepening morass of disunity abdicated at the most crucial moment.

The council had given ample evidence in 1921 that its leadership could be decisive. Its "coordinating council" had provided unity and direction to Progressive efforts during that important campaign. The council had been so successful in this role that enemies of the farmers' political movement bitterly attacked it as the fountainhead of Progressive wisdom.[24] The abolition of the coordinating council in 1923 was applauded by the *U.F.A.* as the "correction of a mistake,"[25] and the *Guide* admitted that now "the control of policy was wholly in the hands of the people."[26] The *Guide* hinted, however, that the wisdom of the move was open to question, for the work of the coordinating council had shown "the need for an agency of the kind." The provincial organizations spoke with many voices; it was no longer possible to talk of a Progressive policy.

Political leaders among the Progressives in the House of Commons were not impressive, and the movement soon showed the demoralizing effects of inadequate leadership. The farmers' representatives in Ottawa, with a few exceptions,[27] were not able parlia-

[22] ". . . the point was eventually reached where it became necessary for the Council either to be re-organized without the commercial companies or cease to take part in election politics. This, I think, is the real reason for the resolution adopted by the Council in March, 1923." Files of the Canadian Council of Agriculture, J. W. Ward (secretary) to William Hirth, June 21, 1924. The author is indebted to Professor W. L. Morton for this interpretation.

[23] *Grain Growers' Guide*, 16:3 (April 4, 1923).

[24] *Farmers' Advocate*, 57:1524 (February 25, 1922).

[25] ". . . its action has made very clear that it will not undertake to interfere with the prerogative of the Provincial organizations in the operation of their political machinery." *U.F.A.*, 2:1 (May 15, 1923).

[26] *Grain Growers' Guide*, 16:5 (April 11, 1923).

[27] These, interestingly enough, were almost all in the U.F.A. group that survived the debacle of 1925. They continued to serve with distinction until the U.F.A. collapse in 1935.

mentarians. The Progressive group in the House suffered from inexperience and inability, but chiefly from the latter. Political observers were deeply critical of the intellectual quality of Progressive representatives. "They never enter the parliamentary library, they will not study blue books, Hansards, or reports, and they will not bother to prepare decent speeches" was the harsh judgment of a journal usually sympathetic to the farmers' cause.[28]

The resignation of T. A. Crerar as parliamentary leader in November 1922 was a sharp blow to the prospects of the new party. His parliamentary experience and his ability were unmatched among the Progressives. With the plea that the declining fortunes of the United Grain Growers, Limited, demanded his full attention, Crerar stepped down from the position he had filled with dignity and restraint. The departure from active leadership of the man so long identified with the farmers' movement in politics was greeted with a universal recognition that the Progressives had lost their most experienced spokesman and their ablest champion in the House of Commons.

The retiring leader made it clear that he disapproved the growing tendency to make the farmers' movement into a class and sectional group. Many considered the resignation as a victory for the Wood "group government" wing of the movement, but the new parliamentary leader, Robert Forke, denied this and assured the House that the change in leadership meant no change in policies.[29] H. W. Wood, on the other hand, met the resignation with the curt statement that Crerar did not believe "in the organization of the people. His proposed legislation is all at the top, none at the bottom. It is political autocracy, as opposed to political democracy."[30] Perhaps the Alberta chieftain was stung by Crerar's parting shot that cooperation with the Liberals in Parliament was opposed by reactionary Liberals and by certain Progressives who were more concerned with the interest of their class or group than with the national welfare.[31]

The struggle for parliamentary leadership following Crerar's

[28] *Canadian Forum,* 3:198 (April 1923).
[29] House of Commons *Debates*, Session 1923, 1:35.
[30] *Canadian Annual Review of Public Affairs*, 1922, 232.
[31] *Ibid.*, 230.

resignation revealed the paucity of trained political leadership in the farmers' ranks. The potential leaders were quickly reduced to two: Robert Forke of Manitoba and J. T. Shaw of Alberta. After a rather bitter struggle involving an effort to reorganize along "broadened out" lines and an incipient revolt of eight of the U.F.A. members, the party caucus finally agreed on the amiable but unaggressive Forke to inherit the Crerar mantle.

That all was not well in the farmers' political movement was increasingly evident. A growing dissatisfaction with the "right wing" leadership of Robert Forke further alienated the U.F.A. element.[32] A grim warning of the coming debacle was furnished to Progressives as their half-hearted or disgruntled colleagues fled to the security of the Liberal party. The defection of J. T. Shaw in November 1922 attracted widespread attention, and the flight to liberalism of W. J. Hammill of Muskoka and Joseph Binnette of Prescott in December was conclusive evidence to many critics that the farmers' movement was fast disintegrating.

The increasing difficulties of the farmers in politics reflected a disastrous growth of political apathy on the part of the farmers themselves. The dangerous tendency of looking upon the good fight as won once the election was over was common among the agrarian crusaders. The familiar pattern of enthusiasm, apathy, and disintegration that had marked so many of the farmers' movements south of the forty-ninth parallel now appeared in the Progressive movement. Farmers' societies experienced a sharp decline in membership as the excitement of the 1921 election wore off. The evangelistic enthusiasm of the election had pushed membership figures to unprecedented heights, but the creeping paralysis of apathy soon attacked these vital nerve centers of the whole farmers' program, and the United Farmers organizations withered away during 1922 and 1923. The S.G.G.A. dropped from 40,000 to 15,000 members in 1922; the U.F.A. membership fell off from 38,000 to 15,000, and the U.F.M. declined to 11,402.[33] The financial strength of the societies also took a similar drop during the post-election years. By 1923 the societies were often operating

[32] This doctrinaire wing included Agnes Macphail of Southeast Grey (Ontario).

[33] *Canadian Annual Review of Public Affairs*, 1922, 227, 742, 828; *Grain Growers' Guide*, 16:13 (January 17, 1923).

with deficits, and the *Guide* feared that the farmers' organizations were in grave danger.[34]

If zeal lagged in the hustings, it also flagged in Parliament, where many of the farmers' representatives displayed a remarkable reversal of attitude. The fiery denunciations of a corrupt and venal House gave way to more restrained judgments, often to acquiescence in the ways of politics.[35] Critics accused the farmers of falling victim to the same demoralizing influences in political life they had condemned so long. The declining zeal for the agrarian crusade placed the growing divisions in the farmers' ranks in bold relief.

Underlying the divisions that rent the Progressive movement was the "broadening out" controversy. This clash over the fundamental nature of the farmers' movement in politics had been divisive from the earliest days of political action. As the years passed, the lines of conflict grew rigid and the struggle was intensified. Each group presented its arguments with force and skill, and farmers were often confused by the conflicting claims of their leaders.

On one judgment all were agreed—the Progressive movement was doomed unless some answer could be found to this perplexing question. The Alberta group uncompromisingly defended the "group movement" concept and adamantly opposed any "broadening out" of the farmers' party. H. W. Wood argued with the force of logic that the movement faced disintegration if the "scattering out" policy, as he chose to call it, was adopted. The Manitoba group, on the other hand, replied with equal vigor that the "class consciousness" of the Wood doctrine was dangerous to the farmers' movement and inimical to the welfare of Canada.

The complete collapse of the United Farmers of Ontario government in June 1923 and the acrimonious controversy that accompanied it were pointed lessons to both western factions. The stunning defeat and sudden liquidation of the Drury government was a harbinger of future events. Discerning Progressives clearly recognized the handwriting on the wall.

[34] *Canadian Annual Review of Public Affairs*, 1922, 227; *Grain Growers' Guide*, 16:23 (January 24, 1923); *ibid.*, 15:5 (October 18, 1922).

[35] See, for example, Alfred Speakman's judgment of the House as the scene of "a high sense of public duty and desire to do what is right." House of Commons *Debates*, Session 1922, 3:2886.

The *Grain Growers' Guide* wrestled valiantly with the problem of finding a workable compromise which might reconcile the increasingly hostile wings of the movement. If urban votes were essential, suggested the *Guide*, then the solution used in Winnipeg and Calgary might be adopted. In Calgary, for example, the U.F.A. had supported nonmembers for office "who were not even asked to subscribe to the farmers' platform."[36] This was not a "broadening out" policy, insisted the *Guide*, but a mutual support of democratic methods where "like-mindedness" existed. Obviously, however, this like-mindedness would be hard to find. The next suggestion from the prairie journal was less helpful. Perhaps, thought the *Guide*, there had been too much emphasis on the urban vote. There was "precious little evidence" that the urban voters desired "to organize democratically as the farmers were organized."[37] The *Guide* had reached an impasse and was reduced to admitting that "the problem looked at in the old way is insoluble," but hopefully predicted that the way out might be found in the "very principle upon which government in Canada is founded."[38] The earnest searching of Progressive hearts that accompanied the controversy failed to produce a solution, and the farmers faced a darkening political future.

The dramatic victory of 1921 which had launched the National Progressive party on its federal career pointed up the sweeping changes in provincial politics that were taking place during the postwar years. Political action by Canadian farmers proved to be more potent and more lasting in western capitals than it was in Ottawa. Farmer politicians early turned to provincial politics to oust the older parties from power and to assume the reins of government for themselves. This was natural in spite of the friendly attitude of the provincial Liberal governments on the prairies. The moral crusade upon which the farmers had embarked, the political aspirations of politically ambitious leaders, and the realities of the politico-economic system in which they sought power forced the move.

The abuses of the party machines at all levels of government

[36] *Grain Growers' Guide*, 15:5 (August 23, 1922).
[37] *Ibid.*, 15:5 (September 3, 1922).
[38] *Ibid.*, 15:5 (August 30, 1922).

aroused the moral indignation of the grain growers. The revolt against partyism on the prairies knew no distinction between provincial and federal politics. This force alone, however, would not have been sufficient to compel the farmers to attempt the defeat of friendly provincial governments. The close relationship so often observed between provincial and federal parties aroused the Progressives' ire and moved them to destroy the roots which federal parties sent down into provincial organizations.[39] There was always the fear that Liberal politicians, however great their professions of loyalty, might betray the farmers' cause in the larger interests of their federal party. These close ties were keenly resented and were a prime cause for the farmers' entry into provincial politics.

The fundamental impulse for political action that accompanied the commercialization of agriculture demanded the creation of farmers' governments on the provincial level. Many solutions to marketing and credit problems were within provincial jurisdiction, and many farmers believed that only through their own governments could adequate legislation be secured. Moral indignation, political realities, and economic necessity were the forces which drove the agrarian revolt into provincial politics, just as the same forces had compelled the American agrarian movements to look to their state capitals for relief. The move into provincial politics, however, seriously embarrassed the National Progressive party as a source of further division and disharmony.

The political conditions that faced the farmers' provincial ventures varied widely, which helps to explain the differing fate of the farmers' governments on the prairies. The urge to enter politics was more powerful in Alberta than in any other province, and the venture there was more lasting. The forces that impelled the U.F.A. into politics have been examined in Chapter VI; it remains to point out here that the farmers' party in Alberta rushed into a political vacuum created by the virtual collapse of the Conservative party and by the lingering but mortal illness that had seized the Liberal machine. The poverty of leadership of the older parties gave the farmers an unusual advantage, for they alone pos-

[39] W. L. Morton, "The Western Progressive Movement, 1919–1921," *Proceedings of the Canadian Historical Association*, 1946, 49.

sessed a Henry Wise Wood. Though Wood consistently refused political honors for himself, his wisdom and experience were ever behind the scenes prompting the government in Edmonton.

Alberta farmers first entered politics through an organization separate from their powerful U.F.A. This was the short-lived U.F.A. Political Association, of which Oran L. McPherson was president.[40] Locals in the U.F.A. were not forced to enter this association and they remained in good standing in the parent society even though they eschewed politics. This was an obvious compromise forced by the reluctance of Wood and other key members of the U.F.A. executive to accept the principle of political action. It proved completely unsatisfactory and quickly raised serious question of relationship to the parent organization. The voice of the new association, the *Western Independent,* soon displayed an independence that embarrassed the U.F.A. When the paper called for the creation of a farmers' "political machine,"[41] the time had come for the U.F.A. to act.

The relationship of the two organizations came under full review in the 1920 U.F.A. convention. Wood vigorously attacked the new organization, claiming that it had attempted to dictate "what he should say." He felt that a separate political organization was unnecessary because its work could be accomplished more effectively and at less expense through the U.F.A. itself. In its stead he proposed the "group government" concept of political action which would place the U.F.A. in politics as an economic organization rather than a farmers' political party.[42] Feeling ran high as the farmers battled over the issue. A. C. Muir led a bitter attack against President Wood, and J. W. Leedy and Wood parted company over the question.[43] The convention sustained its presi-

[40] McPherson possessed previous experience in farmers' societies in Illinois. He had been trained in Shurtleff College and the University of Illinois. He later served as speaker of the Alberta legislative assembly.

[41] *Western Independent,* 1:8 (November 12, 1919). This paper was the successor to the *Alberta Non-Partisan.* The ubiquitous William Irvine remained as editor.

[42] "The political party is not an organization. If the farmers should inaugurate a straight farmers' party and adopt a straight farmers' platform, they could not carry on democratically because the political party structure can not implement democratic political action." H. W. Wood, "The Group versus the Party," *Grain Growers' Guide,* 15:17 (September 6, 1922).

[43] Muir revived the old charge aired some time earlier by P. P. Woodbridge, one-time secretary of the U.F.A., that the Alberta society had centralized control in the hands of an autocratic "two or three" men.

dent in the voting, however, and the U.F.A. Political Association was dissolved.

The U.F.A.'s venture into provincial politics enjoyed unprecedented success. The farmers swept into power in 1921 and retained their position in Edmonton for fourteen years. Herbert Greenfield, though he was not Wood's first choice, was selected to head the new government. In 1925, however, the farmer-premier stepped down in favor of J. E. Brownlee, who had been Wood's choice in 1921 and was the only lawyer in the U.F.A. government.[44]

The political policy of the new government followed the pattern cut out for it during the decade of agrarian agitation which preceded the seizure of power. The dire predictions that the farmers would lead Alberta into financial ruin and into political sovietism proved false, for the farmers' government quickly settled down to provide the province with a thoroughly competent administration. The Greenfield government carried the heavy burden of the severe agricultural depression which reduced provincial incomes and caused widespread distress. Low prices and severe droughts imposed the necessity of seed grain acts and relief acts upon the U.F.A. government. This in turn compelled the Greenfield cabinet to adopt severe economies and led the government into an increasingly conservative administration. The premier denounced the growing "paternalism" and declared, "We have overstepped the limit in recent years and our people have learned to lean unduly on the government for assistance."[45] This was a far cry from the radicalism expected of the farmers.

The swing away from the more radical element in the U.F.A. was dramatically illustrated by the internecine struggle over banking and credit legislation. The radical wing, led by George Bevington, strongly urged the nationalization of banking and credit facilities. The cooperative credit societies, which were rapidly expanding their service of rural credits, were inadequate in Bevington's view. He emphasized the social aspects of credit control. The

[44] Greenfield admitted, "It has become increasingly evident that there is not the full accord between the supporters of the government and myself which is essential to the successful combination of any administration." *Grain Growers' Guide*, 18:40 (December 2, 1925).

[45] *Canadian Annual Review of Public Affairs*, 1923, 737.

basis of credit, he maintained, was "the number, intelligence and industry of the people, plus the capital equipment and natural resources of the state."[46] The function of credit he defined as ultimately social, for the maintenance of production and distribution were to be valued only as they promoted human well-being. It naturally followed, argued Bevington, that the only agency qualified to control the flow of credit was the government. Bevington and his colleagues preached a gospel of social credit long before the average Albertan had heard of William Aberhart or Major Douglas.

During 1922 and 1923 the U.F.A. followed Bevington's lead by adopting resolutions advocating the creation of a provincial bank. U.F.A. representatives in Ottawa were urged to aid in establishing a federal loan department which would issue full legal tender notes from time to time and provide loans to farmers through provincial agencies. Pressed by the U.F.A. conventions, the Greenfield government commissioned Dr. D. A. MacGibbon of the provincial university to investigate the operation of the Dominion Bank Act in Alberta. The resulting report discouraged the project of a provincial bank as impractical and costly, since the "valuable end of the business" was already controlled by the chartered banks.[47] This dampened the enthusiasm of the government so much that it dropped the project and refused to enact any legislation despite U.F.A. pressure.

The refusal of the Greenfield government to act on the provincial bank and credit schemes precipitated a struggle within the U.F.A. The opponents of easy credits and soft money damned the Bevington scheme with unrestrained enthusiasm. Henry Wise Wood denounced the plan as "dangerous" and "liable to lead to disaster." He warned that similar demands had wrecked Farmers' organizations in the States. The *Grain Growers' Guide* condemned the program as a threat to the whole farmers' movement, and published a series of articles by Harald S. Patton to offset the heterodox financial schemes that seemed to thrive in the western atmos-

[46] House of Commons *Journals*, Session 1923, Appendix No. 2, "Proceedings of the Select Standing Committee of Banking and Commerce," 127. Hereafter cited as House of Commons *Journals*, Session 1923, Appendix 2.

[47] The report was unpublished, but pertinent excerpts may be found in House of Commons *Journals*, Session 1923, Appendix 2, 182.

phere. The *Guide* joined Wood in citing the frequent concern of American farmers' movements in currency and credit manipulation, and warned the U.F.A. that it was pursuing the same "will-of-the-wisp."[48]

The issue came to a head in the U.F.A. convention of 1924 when a Bevington resolution was defeated. This led to a brief but unsuccessful effort to challenge Wood's power. Sharp words were exchanged and tempers flared as Wood led the move to crush the soft-money group. Bevington's influence in the U.F.A. steadily declined as the more conservative monetary views of Wood prevailed. Interest in currency and credit manipulation remained, however, and in the following year the convention endorsed a "revolutionary" scheme providing for the issue of currency by the dominion government on the security of provincial bonds.[49]

In Manitoba the decision to enter provincial politics came later and aroused greater opposition than in Alberta. Though the Liberal Norris government showed every desire to meet the farmers' demands, the rank and file of the U.F.M. were restless and urged their leaders to act. In 1920 the pressure from below was so great that the executive allowed the locals to take political action if they desired, but made it clear that the Norris government had "redeemed its pre-election pledges more than any other government on record."[50] The U.F.M. refused to take part in the election of that year. The nine farmer representatives who were elected had been nominated and supported by local constituencies, not by the central organization.[51]

The result was not a happy one. Candidates with partisan records engineered conventions and secured endorsements as "farmers' candidates." The genuine farmer candidates resented the lack of aid from the central office, and discontent was general through-

[48] *Grain Growers' Guide*, 16:5 (January 24, 1923); *ibid.*, 14:5 (February 2, 1921); *ibid.*, 17:5 (January 23, 1924).

[49] Recognizing the strong feeling for soft money in the West, the *Guide* admitted, "The *Guide* could make itself very popular with some of its readers if we were to endorse . . . the many schemes that are being put forward to get cheap money." (18:5, March 4, 1925.) The scheme to pledge provincial bonds for an issue of dominion notes may be found in Robert Gardiner, "Memorandum on Rural Credits" (House of Commons *Journals*, Session 1923, Appendix 3, 1x).

[50] *Grain Growers' Guide*, 15:20 (July 14, 1920).

[51] *Ibid.*, 15:6 (July 7, 1920) .

out the organization. The upshot was that the U.F.M. decided to take the plunge into politics.

The uneasy alliance that sustained the Norris government collapsed in March 1922, and the farmers faced their first real test of strength in Manitoba.[52] The typically Progressive platform with which the U.F.M. entered the campaign had come up from the locals and was only adopted by the convention in January 1922.[53] Clad in the armor of reform and denouncing the "infamous" alliance of the Norris government with the federal Liberal party, the U.F.M. entered the lists. Premier Norris, who had been praised as the "true friend of the farmer" two years earlier, was now bitterly assailed. His eulogy of the Mackenzie King administration aroused keen resentment and offended the deep-seated Progressive bias against such a relationship. It reinforced the farmers' conviction that the powerful federal parties must be weakened by destroying their provincial foundations.

The U.F.M. won a sweeping victory just as the U.F.A. had earlier captured Alberta. The farmers were now ensconced in two of the three prairie capitals. The new government in Manitoba called President John Bracken of the Manitoba Agricultural College to serve as premier. The new premier and all but one of his cabinet lacked legislative experience. The Manitoba government carried the burden of inexperience so characteristic of farmers' governments everywhere on the continent.

The new administration inherited serious financial difficulties and the provincial government soon faced deficits. The rural credits program was operating with serious losses. Even the *Guide* admitted that years would be required "to straighten out the blunders made in the operation of the Rural Credit societies."[54] The most serious embarrassment for the Bracken government came with the rejection of the wheat board bill in 1923.[55] U.F.M. representatives in the legislative assembly did not take kindly to party discipline and often placed the life of their government in danger by voting with the opposition. By 1924, however, they had settled

[52] The Norris government relied upon the support of labor and farmer groups in the House as a result of the election of 1920.

[53] *Grain Growers' Guide,* 15:5 (February 1, 1922).

[54] *Ibid.,* 16:5 (February 21, 1923).

[55] See page 174.

down, so much so that the life of the government was threatened only once in that year.[56]

The farmers had been in power only a short time when the U.F.M. began to shift its emphasis away from political action. President C. H. Burnell warned in 1923, "We must be careful lest we come to regard our political activities as the main function of our association and politics as the panacea of all our ills."[57] By 1925 the major emphasis of the society was again on the economic reforms sponsored by the organized farmers. Faith in political action was declining.[58]

The foundations of the U.F.M. government were imperiled, moreover, by the steady decline in the society's membership.[59] A rival farm organization, the District Builders, appeared in southern Manitoba to challenge the supremacy of the U.F.M., and apathy also leveled its toll against the society. In 1925 the *Guide* expressed grave fears that the U.F.M. "was moving toward extinction."[60] The farmers' provincial party thus faced the same declining strength in its local organizations and suffered the same loss of enthusiasm among the rank and file that threatened the National Progressive party. The result was the same for both in Manitoba—the farmer politicians drifted into the Liberal-Progressive alliance and ultimately found shelter in the Liberal party.

In Saskatchewan political conditions denied the farmers the victories they had won in the other western provinces. Though the Conservative party was moribund, the Liberal party possessed unusual vitality. It boasted a shrewd and able leadership and commanded one of the most effective political machines on the continent. Against this combination the Progressives made slight headway.

The Liberal government in Saskatchewan was a grain growers' government in all but name. A majority of the legislative assembly were members of the S.G.G.A. and Premier Charles A. Dunning

[56] The U.F.M. government had twenty-five seats while Liberals had seven and Labor had seven. In spite of this, the Bracken government survived several divisions by a one-vote plurality.

[57] *Grain Growers' Guide*, 16:4 (January 17, 1923).

[58] "Some Pertinent Facts about the U.F.M.," U.F.M. pamphlet, 1925.

[59] *Grain Growers' Guide*, 17:8 (January 16, 1924). U.F.M. membership declined to 5881.

[60] *Ibid.*, 18:5 (January 14, 1925).

had come up through the farmers' organizations. His government met with alacrity every legitimate demand made by the farmers. Though Dunning approved the New National Policy, he steadfastly denied the necessity of provincial political action by the S.G.G.A.[61] With eloquence and ability he preached a doctrine of the separation of federal and provincial parties, and remained ever attentive to the grain growers' wishes.

The pressure to enter politics increased as the Progressive movement gained momentum. The executive of the S.G.G.A. resisted the trend and fought the rising passion for politics. The Saskatchewan leaders marshaled their arguments skillfully and kept the revolt in check. Whenever the demand for provincial political action grew irresistible, the executive simply referred the matter to the locals for action. The result was an occasional effort to challenge the powerful Liberal machine. Most of the grain growers shared the conviction of their leaders that without firing a shot they had captured the field, for they controlled the legislature and possessed the cabinet.

In spite of this, the demand for independent political action assumed serious proportions in 1922. In that year the advocates of political action found unexpected support from J. B. Musselman, who reversed his stand of several years and joined John Maharg and George Langley in demanding an entry into provincial politics. Under their tutelage, the S.G.G.A. decided to enter the provincial arena and authorized the central board to create a committee to assist the constituencies in organizing for political action.[62] Shortly after this significant break with the past, President Maharg issued a circular letter to all S.G.G.A. locals urging them to eliminate all government influences and ordering them to remove all government employees from their election organizations.[63]

The movement was clearly assuming menacing strength. There were grave fears in government circles that this revolt, unlike the effort of the independents led by Harris Turner in 1921, could not be headed off by another snap election. Premier Dunning fought the movement with skill and resourcefulness. He pleaded

[61] *Ibid.*, 13:19 (August 4, 1920). [62] *Ibid.*, 15:3 (February 22, 1922).
[63] *Canadian Annual Review of Public Affairs*, 1922, 799.

with the farmers' organization not to establish a "political test of membership," and insisted that the "bulk of the members of the S.G.G.A. are Liberals." He made his own position perfectly clear with the ringing challenge, "If you want a man to fight the Grain Growers' Association, go out and look for someone else." [64]

Time and skillful political jockeying soon removed the threat. The resignations of J. B. Musselman and John Maharg to accept positions of responsibility with the Saskatchewan Cooperative Elevator Company cut the heart out of the S.G.G.A. political ambitions. These two men had been the leading spirits in the political revolt, and their move to the cooperative company placed them in an organization closely allied to the government and dependent on it for support.

The rapid growth of the Farmers' Union in Saskatchewan also weakened the S.G.G.A. at this crucial juncture. This society, borrowed from the United States like the Grange, the Alliance, the Equity, and the Nonpartisan League before it, swept through Saskatchewan in 1921. It quickly gained a following among farmers who were dissatisfied with the S.G.G.A. Led by the fiery L. P. McNamee, the Farmers' Union cut a widening swath through the province until in July 1926 it dramatically joined forces with the S.G.G.A. to form the United Farmers of Western Canada, Saskatchewan Branch.

A Provincial Progressive Association was launched in late 1923. Harris Turner, leader of the opposition in the Saskatchewan Assembly and editor of the *Progressive,* headed the new party. Though Turner denounced the "menacing machine" that supported Dunning, and stigmatized the Saskatchewan Liberal party as a "donkey engine for the government in Ottawa," his appeals failed to strike fire. The *Grain Growers' Guide* criticized the Saskatchewan Progressives for creating the same relationship with the federal Progressive party they condemned in the Liberals.[65] The political movement failed completely, and the incipient revolt against the Dunning government collapsed. In the 1924 convention the S.G.G.A. repudiated its 1922 action and withdrew from

[64] *Grain Growers' Guide,* 15:6 (April 12, 1922).

[65] The *Guide* warned the Saskatchewan party that the separation of federal and provincial parties had "an historic background, and should not be lightly abandoned." (16:5, November 7, 1923.)

provincial politics altogether.[66] The following year the Dunning government was returned to power in an easy victory. The Saskatchewan Liberal government was thus the only prairie government to survive the farmers' revolt. It performed this feat by bending to the farmers' will and by severing its ties with the federal party.

In Ottawa, meanwhile, the National Progressives sought to apply the political power won in 1921 to the solution of agricultural problems. Foremost on the list of farmer demands was the plea for a reconstructed wheat board to check the disastrous decline in prices that followed the war. The National Wheat Marketing Agency created by the new Parliament lacked some of the powers of the 1919 board, but it was given all the powers within the jurisdiction of Parliament. Concurrent legislation by the provinces was necessary to provide additional powers and privileges to enable the new board to function with the effectiveness of the 1919 board. Prospects seemed bright that the grain growers would be successful in their long fight for a wheat board, for they controlled all three of the prairie governments.

Provincial action followed quickly. The U.F.A. legislature at once pushed through enabling legislation and the friendly Liberal government in Saskatchewan soon followed suit. The moment seemed propitious to launch the new wheat board, and the Canadian Council of Agriculture optimistically expressed its willingness early in August to cooperate in establishing a board to be in operation for the 1922 crop.[67] Within two weeks, however, the whole program received a stunning blow when James Stewart and F. W. Ridell of the 1919 board declined to serve on the new board. Their refusals were followed by declinations by H. W. Wood, C. Rice-Jones, and others competent to handle the responsibility. On August 14 Premiers Dunning and Greenfield announced their intention of abandoning further efforts to establish a board for that crop year owing to the impossibility of finding competent men.[68] Hopes were high, however, that during the following year this problem would be solved.

[66] *Ibid.*, 17:5 (January 30, 1924).
[67] *Ibid.*, 15:3 (August 9, 1922).
[68] *Ibid.*, (August 23, 1922).

The Manitoba legislature shattered these hopes and struck the wheat board plan a fatal blow in April 1923 by defeating the measure. Six U.F.M. members deserted their government on this crucial issue and voted with the opposition to defeat the bill, twenty-four to twenty-one. This surprising turn of events aroused the keen resentment of Manitoba farmers and placed the two western provinces in an extremely difficult position.[69] Late in June the western premiers again issued a joint statement, this time revealing that the creation of a wheat board was now an impossibility. The *Grain Growers' Guide* greeted this last announcement with the death notice of the wheat board scheme — "The Wheat Board is now dead."[70] In July, however, the dynamic evangelist of agricultural cooperation, Aaron Sapiro, appeared in western Canada. The prairies were soon aflame with wheat pool schemes, and the farmers turned to the collective risk-bearing plan of the pool rather than to the government wheat board. Ironically, the farmers who now possessed the political power to realize their dream of a compulsory wheat board, were forced to turn to the voluntary pooling scheme.

Progressives in Parliament fought other and more successful battles for their western constituents. They successfully demanded an inquiry into the grain trade in 1923, and won a very important struggle to restore the Crow's Nest Pass railway freight rates on wheat and flour.[71] They also fought the high freight rates and the shipping monopoly on the Great Lakes, but with less success. The western farmers firmly believed that many of their transportation and marketing problems could be reduced if the Hudson Bay Railway was completed. Their representatives in Parliament therefore continually pressed for the completion of the line. They also demanded the return of the natural resources from dominion custody to the western provinces but, as in the railway issue, they were always checked in these ambitions, and success was not to come until the National Progressive party had vanished.

Progressive members of Parliament took a keen interest in

[69] The *Winnipeg Tribune* attributed the defeat to the fact that "Mr. Bracken was forced into an invidious position by politicians more experienced and skillful than himself." (May 5, 1923.)

[70] *Grain Growers' Guide*, 16:5 (June 27, 1923).

[71] "The powerful lobby maintained by the C.P.R. at Ottawa went down to defeat before the Progressives." *Ibid.*, 15:5 (July 5, 1922).

banking and credit legislation. The western representatives boldly denounced the Canadian "banking monopoly" and lashed out at the "pirates of St. James Street." They did not hesitate to announce their firm conviction that the West suffered from the existing system.[72] The dramatic failure of the Home Bank and its many western branches in 1923 brought immediate demands for inquiries into the banking system. Progressives offered no less than six resolutions in the House of Commons calling for a national banking system. When the Canadian Banking Act came before the House for revision, Progressives fought to lower the legal interest rates and demanded that there be periodic inspection of all Canadian banks.

The question of rural credits was always a lively one with the farmers' representatives in Ottawa. They considered it one of the major problems of the country. In 1923 they marshaled their arguments before two House committees, one on banking and commerce and the other on agricultural conditions. It was apparent from the proceedings of both committees that western Progressives shared the farmers' conviction that prairie agriculture bore a heavy and unjust burden. The farmers were fast being reduced to farm tenancy and serfdom,[73] and their representatives were determined to use their political power to avert this national disaster. They looked upon the banking and credit systems of Canada as peculiarly ill-adapted to prairie agriculture, designed as they were to meet the needs of the industrial and commercial system.[74] The remedy, they believed, lay in longer term loans at lower interest rates provided by the government. Not until the Liberal government faced certain defeat without Progressive support were the farmers able to secure a rural credits' bill. By the time the Senate had finished with it, however, it was a pallid shadow of what the farmers really wanted.

The Progressive revolt in the West was more than an effort to secure favorable economic legislation; it was also a moral revolt demanding political reform. Numerous resolutions introduced in

[72] "The banking business of Canada is now in the hands of the directors of 17 banks, and through the Bankers Association, the directors of three or four different banks are able to dictate the financial policy of this country." George G. Coote, House of Commons *Debates*, Session 1923, 5:4073.

[73] House of Commons *Journals*, Session 1923, Appendix 3, 1099, 1100.

[74] *Ibid.*, lix.

the House by Progressives bear testimony of the democratic faith that motivated the grain growers in politics. The western delegation defended its use of the recall in the face of bitter criticism. O. R. Gould first bore the brunt of the attack when he admitted that 40 percent of the voters in his constituency could demand his recall if he refused "to do what the committee asks me to do on the floor of this House."[75] This close tie with the farmers' organizations led to accusations that the Progressives were nothing more than "delegates of their associations" or, as Henry Spencer was called somewhat later, "gramaphones of the United Farmers of Alberta."[76] The House rejected the move to disqualify those who signed recall agreements, but the recall idea remained an unpopular innovation.

A further evidence of the reform nature of progressivism was the effort to exclude from the cabinet any member of Parliament who held a directorship in a bank, trust, or insurance company, public utility, or transport corporation. The resolution was soundly defeated as Liberals and Conservatives turned on the "radicals" to overwhelm it, 142 to 59. A resolution by William Irvine sought to alter the parliamentary practice of accepting any vote as a vote of confidence. The Irvine measure was the familiar U.F.A. program already adopted in Alberta to end cabinet domination. The resolution declared that "the defeat of a government measure should not be considered as a sufficient reason for the resignation of the government, unless followed by a vote of lack of confidence."[77] The resolution brought sharp charges of "undemocratic," "un-British," "American inspired," and "Bolshevism." The farmers, of course, sought to free the individual member of Parliament to consider each issue on its merits, thus placing greater responsibility upon the Parliament.[78] Again the older parties solidly opposed the Progressives as the "balance of power" operated in reverse.

The chief cry of the revolt in western Canada was for tariff reform and the reality of Progressive influence in the House can be measured by the budgets brought down by the King government.

[75] House of Commons *Debates*, Session 1920, 1:1181; *ibid.*, 2:2031.
[76] *Ibid.*, Session 1923, 5:4058.
[77] *Ibid.*, Session 1923, 1:208.
[78] *U.F.A.*, 2:4 (March 1, 1923).

In 1922 and 1923 the budgets made slight concessions to the low-tariff bloc. The Progressives received these budgets coolly and continued the good fight against Protection.[79] The declining strength of the Liberal party in the House of Commons, however, forced a change in government policy. The defeat of Liberal candidates in by-elections and the retirement of Sir Lomer Gouin and W. S. Fielding forced the prime minister to "steer West" in his tariff policy to placate Progressives, thus preventing the collapse of his government. The Progressives enthusiastically greeted the tariff reductions of the 1924 budget and confidently looked to a future in which the government would sponsor Progressive legislation. Conservatives denounced the Liberal move as "a surrender to the Progressives," and stigmatized the King cabinet as "a cabinet in custody." The prime minister denied this, of course, and insisted that the Liberal party was fulfilling its pledge of the 1919 platform. He began, however, to emphasize the threat of "Toryism" and publicly deplored the division between Progressives and Liberals when "our policies are the same." Behind the scenes, the prime minister reopened negotiations with certain Progressives looking to a coalition. The Progressive attitude was clearly demonstrated when Robert Forke joined Liberals to oppose the Woodsworth amendment, which sought to place the "necessaries of life" under free trade. Forke explained this apparent desertion of Progressive doctrine by characterizing the amendment as a "mere gesture without meaning."

The *rapprochement* of Liberals and Progressives finally split the Progressive group into warring factions and the divisive forces that had worked so long to undermine the Progressive movement were now apparent. The refusal to accept political responsibility, the unwise devotion to the balance of power strategy, the lukewarm loyalty of Progressives who remained Liberals at heart, the costly mistakes of inexperienced leaders, and the growing apathy of the farmers to their own political crusade had destroyed the foundations of Progressive political strength.

In June a group of six notified Forke that they would no longer be guilty of "playing the Liberal game." Four more dissatisfied

[79] "We object to having to live indefinitely under a policy from which we receive no benefit and for which we have to pay the price." Alfred Speakman, House of Commons *Debates*, Session 1922, 3:2884.

Progressives soon joined the revolt.[80] This "Ginger Group" looked upon the *rapprochement* as a frank admission by their colleagues that they were "seeking a way of escape by taking steps toward the adoption of the old party system."[81] The disintegration of the movement was well under way. It was a touch of irony that at the very moment when Progressive bargaining power was at its zenith the movement went to pieces, thus rejecting by default the prize it had sought so long.

The serious divisions in the farmers' ranks struck a death blow to Progressive hopes of molding Liberal policy in the House. In July the Progressive parliamentary party announced it had resolved its differences and a "definite basis for cooperation of all members had been outlined and approved."[82] The new unity was more apparent than real, and the mortal illness of the party could not be concealed.[83]

The obvious weakness of the Progressive party led the prime minister to believe that an appeal to the people would end the growing instability in the House of Commons by electing Liberals rather than Progressives. This would give the government a clear-cut majority and end the intolerable uncertainty of not knowing from day to day whether the government would have a majority. The Progressives, badly divided, poorly led, and now a party without a dominion-wide following, had fumbled their great opportunity as a national party. Their course was nearly run, and it was obvious that the farmers' party in Canada was destined to go the way of all agrarian parties on this continent.

[80] These dissenters were H. N. Campbell, Robert Gardiner, E. J. Garland, D. M. Kennedy, H. E. Spencer, Agnes Macphail, W. C. Good, George G. Coote, Preston Elliott, and W. J. Ward.

[81] *U.F.A.*, 2:3 (December 15, 1924).

[82] *Grain Growers' Guide*, 18:3 (July 8, 1925).

[83] "The division," admitted the *Guide*, "has caused considerable glee in some of the camps of the enemy." (17:5, July 9, 1924.)

X

The Progressive Heritage

The National Progressive party collapsed in the federal election of 1925. The calm tempo of the rather colorless campaign of that year concealed the momentous effects the election held for Canadian politics. Both Liberals and Conservatives conducted "safe" campaigns as they battled over issues that had been grist for the political mill for a generation. The prime minister denounced the Progressives as "political outlaws" who made the West "ridiculous."[1] Liberals made every effort to capture western votes, for they were essential to a strengthened liberalism in the House of Commons. Promises to bring western men into the cabinet and to complete the Hudson Bay Railway project were held out as bait to Progressives to lure them back to the party of their first love. Appeals to Progressives to drop their "spirit of isolation" and to unite with their "natural friends" in the great struggle against Toryism were Liberal stock in trade in the West. The Conservative campaign, on the other hand, was primarily devised to capture the East. Ontario was reminded that the Liberals had sold out to the West, while in Quebec reaction against the Liberal party gained headway under the Patenaude banner.

The Progressive campaign in 1925 was a pallid transcript of the stirring crusade of 1921. The same issues were present, but the old fire was missing.[2] The tariff, freight rates, the Hudson Bay Railway, an elective Senate—these were familiar Progressive issues, but there were disquieting signs that the movement was faltering. In Ontario there was every indication that the movement would be swept aside, and even in the West the tempo of the Progressive campaign alarmed many sympathetic observers. The

[1] *Manitoba Free Press*, October 12, 1925.

[2] Robert Forke warned his constituents that the Progressive party was no Moses "to lead us out of the present land of bondage to the Promised land." *Manitoba Free Press*, October 17, 1925.

party experienced unexpected difficulty in financing its campaign, and many Progressive candidates were left to face powerful Liberal machines in their constituencies with their own slender resources. The *Guide* was plainly alarmed as it witnessed the once-powerful movement facing the superior organization of the older parties without adequate financial backing. "When Progressive candidates have to finance their own campaign," lamented the editor, "then there will be mighty little Progressivism in the Progressive party."[3]

The financial poverty of the party was a surface indication of a deeper and more serious condition. Except in Alberta, where the U.F.A. maintained a tight discipline and the *esprit de corps* was high, the fervor had departed from the farmers' movement. The election of 1925 came at a time when the wheat industry was fast recovering its lost prosperity. A bad crop in 1924 or 1925 might have revived the crusade, but the West was entering a period of favorable prices and active export which lasted until the Great Depression of 1929. The improving economic situation on the prairies undermined the party of protest. The farmers' energies, moreover, were now turned to the development of the wheat pools, and there was little enthusiasm left for the political crusade.

The Progressive campaign was crippled by the fact that Progressive candidates could offer nothing to the voters in terms of political power. There was little hope that they could win enough seats to become the opposition in the House, and none that a Progressive government could ever be formed. Even a balance of power seemed out of the question. Many voters refused to waste their votes on such a party, especially when western Liberals promised a similar program. The work of Premier Dunning was especially damaging to the Progressive cause in Saskatchewan. His powerful machine convinced thousands that a vote for the disorganized Progressives was a vote wasted.

The mortal weakness of the National Progressive party was obvious to its critics and suspected by its friends. The *Guide* deplored the lack of "educational work" by the party, it criticized the serious lack of organization, and lamented the financial poverty. Two weeks before the election, this organ of prairie Progres-

[3] *Grain Growers' Guide*, 18:5 (October 14, 1925).

sivism grimly warned of the coming debacle: "There is not the same enthusiasm in the country during the present campaign as marked the campaign of four years ago. Progressive supporters must bend all their energies to the work of electing their candidates." [4]

Only in Alberta was there real confidence of victory. There the U.F.A. carried on a vigorous campaign against any return to "party machine government." The issue was a real one to Alberta farmers, and they were determined "to thresh out an overflowing harvest of golden principles . . . and seal and consecrate those principles for the future by the very sincerity and unwavering faith in which we go to the ballot box in this election." [5] There was no loss of zeal among the "organized citizenship" of Alberta!

The election results came as a shock to most Canadians. Party standings were completely reversed in the new Parliament. The Liberals, who had hoped to emerge with additional seats at Progressive expense, suffered staggering losses as people "voted for a change." The prime minister and seven of his cabinet failed of reelection, and Liberal strength in the House of Commons was slashed from 128 to 101 seats. Mr. Meighen's Liberal-Conservative party, on the other hand, nearly doubled its strength.

The National Progressive party was the real victim of the election. It lost two thirds of its membership in the House, and its popular vote fell to less than 285,000. The Ontario delegation melted away under the Conservative wave that swept that province. The eastern wing of the movement was cut from twenty-two to two, and the Progressive movement was reduced to a prairie province group. The election of 1925 thus humbled the Progressive movement to a party of sectional protest as the center of agrarian radicalism shifted to the far West. The *Manitoba Free Press* fashioned the epitaph of the National Progressive party when it regretfully observed, "The Progressives . . . are probably going the way of all third parties in Canada." [6]

Though the election nearly wiped out the Progressives, they held the political balance in the House of Commons more firmly

[4] *Ibid.*, 18:5 (October 14, 1925).

[5] Bert Huffman, "A Harvest of Golden Principles, or a Harvest of Chaff," *U.F.A.*, 4:4 (October 24, 1925).

[6] *Manitoba Free Press*, October 30, 1925.

than ever. This curious paradox resulted from the fact that neither Liberals nor Conservatives possessed a working majority. The election had only made matters worse! The twenty-five Progressives were the key to the new Parliament, and no government could survive without their support.

Western Progressives welcomed this unprecedented development, for now the King government would be forced to pursue "a genuine Liberal policy."[7] There was one fear to temper western enthusiasm — could the western members of Parliament unite their disorganized forces to face this new opportunity? The answer came in the form of a press announcement by Robert Gardiner of the U.F.A. Gardiner confirmed the fears of the most pessimistic when he stated: "The U.F.A. is a distinct parliamentary unit, and is not part of any political party. It is under no obligation to give support to Liberals, Conservatives, Progressives or any other party."[8] The *Grain Growers' Guide* received this news with the prophecy that "what has been known as the Progressive group will not be a unit in the new Parliament."[9] The farmers' journal doubted that the divided party would accomplish very much for the western farmers.

The new Parliament was doomed to an early death. The fate of the Mackenzie King government rested squarely with the farmers' representatives. Conservatives and Liberals frantically bid for Progressive support, and the farmers were the object of devoted attention from both sides of the House. Mr. King forgot his earlier "political outlaw" charge, and wooed the farmers with promises of further tariff reductions, a rural credits bill, and the completion of the Hudson Bay Railway. Conservatives also vied for Progressive affections, but their proposals were coolly received. The Liberals were the successful suitors, for Progressives quite naturally believed that Mackenzie King would give them more progressive legislation than their old enemy, Arthur Meighen. The Progressive caucus, therefore, voted to cooperate with the King cabinet,[10] and the prime minister's decision to carry on the gov-

[7] *Grain Growers' Guide*, 18:5 (November 11, 1925).

[8] *Ibid.*, 18:5 (December 9, 1925).

[9] *Ibid.*

[10] "When we came here at the beginning we entered into a cooperative agreement with the Liberal party, believing it more likely that we would get legislation we wanted." Agnes Macphail, House of Commons *Debates*, Session 1926, 5:5149.

ernment despite his minority position in the House was rewarded with a Progressive alliance. Fruits of cooperation were soon revealed as Progressives rallied to the Liberal cause to defeat a Conservative resolution of "no confidence." A grim forecast of the tense days ahead was discernible in the division, however, for five farmer members bolted their caucus, and voted with the Conservatives.

Cooperation between the two parties depended not only on Liberal concessions, but also upon the ability of the Progressives to preserve a party discipline. If enough Progressives bolted their caucus, the King government would surely fall. Progressive discipline had failed to hold the farmers in line in the preceding Parliament, and the party's present advantage could easily be lost by another collapse of solidarity. The "Ginger Group" prevented an outright alliance, and the cooperation remained on a day-to-day basis.

The customs scandal was the issue which finally ended the cooperation and led to the fall of the King cabinet. The scandal had all the elements of political corruption that the Progressives had condemned for many years. That it ruptured the alliance was inevitable. A special House investigating committee revealed that a gigantic smuggling industry had prospered in the illegal shipment of liquor from Canada into the United States. Even worse, to increase their illicit profits the thrifty smugglers loaded their craft in American harbors with valuable dutiable goods for their return trip. These were then placed on the Canadian market in competition with Canadian goods without paying any customs duties.

The parliamentary committee reported that the smuggling had been possible only through official connivance and by the lax administration of the Customs Department. The vote of censure in the committee was passed on a strict party basis with the only Progressive, D. M. Kennedy, casting the deciding ballot. The charges were soon elaborated to include lurid stories of "pleasure trips" and liquor parties involving members high in the government aboard the *S.S. Margaret.* Stories of "gross immorality" in the civil service in Ottawa were also circulated to discredit the government.

Progressives were shocked by the revelations. They had come to

Ottawa as representatives of a crusade against this very thing. Here was the kind of issue that symbolized the evils their party had set out to crush in Canadian politics. "There is now a concrete opportunity . . . to put our principles into effect, and to strike a real blow for Canada," was the way A. M. Boutiller phrased it.[11] W. T. Lucas called it "the acid test of the sincerity of Progressives."[12] Debate in the House reached fever pitch as Liberals and Conservatives labored to influence Progressive votes. Prime Minister King's defense of the former minister of customs, Jacques Bureau, alienated several Progressives who were shocked that King would go so far in defending his government.

Progressive party discipline was shattered under the impact of the dramatic revelations. "I may frankly state that the hon. member for Brandon [Mr. Forke] has never been able to speak for the Progressives and is not able to speak for them today," declared one Progressive who had decided to bolt the caucus.[13] As the fiery debates continued through the hot days of late June, it became increasingly apparent that the King government was in serious trouble. Party whips worked constantly to prevent a snap vote from upsetting the government. Finally on June 28 came the prime minister's dramatic announcement: "The public interest demands a dissolution of this House of Commons. . . . His Excellency having declined to accept my advice to grant a dissolution, to which I believe under British practice I was entitled, I immediately tendered my resignation."[14]

The Progressives now faced another fateful decision. Should they cooperate with the new Meighen government and thus keep it in power? A Progressive caucus immediately presented a memorandum to Robert Forke, the party leader, approving assistance to the new administration which the prime minister sought to create. The caucus made it clear, however, that the assistance was purely temporary and was only an effort to complete the business before the session. On June 30 nine Progressives joined with Conservatives to sustain the "shadow cabinet" of acting ministers

[11] House of Commons *Debates*, Session 1926, 5:4999.

[12] *Ibid.*, 5:4920. Mr. Lucas referred to Bureau as the "ex-Minister of Customs, under whose administration the bootlegger and smuggler flourished like green bay trees."

[13] M. Campbell (Mackenzie). *Ibid.*, 5:5017.

[14] *Ibid.*, 5:5096.

which Mr. Meighen had set up. Two days later, however, all but three Progressives deserted the government and voted with Mr. King's Liberals.[15] The governor general then granted to Meighen the election he had earlier denied to Mackenzie King.

The Progressives were abused by both parties in the acrimony that followed. Conservatives accused Forke of having pledged his party's support to the Meighen government, but there is no substance to this charge. The Progressive announcement of cooperation was on the same day-to-day basis as the earlier understanding with the Liberals. The Progressives were simply trying to salvage as much as possible from the unproductive session.

The election of 1926 was largely fought over the issues created in the session of 1925. The customs scandal gave the Conservatives a capital issue, but they overplayed their hand. The Liberals found even better campaign material in the "constitutional issue." They were able to use it as an appeal to the same spirit of Canadian nationalism that the Conservatives had exploited so successfully in 1911. Liberal campaign speeches denounced the governor general's intervention as a dangerous threat to Canada's autonomy. Mackenzie King was careful, however, to criticize Mr. Meighen's conduct as unconstitutional, not the governor general's.

The Progressives virtually disappeared in the election of 1926. Most Progressives in Manitoba joined the Liberals to campaign as Liberal-Progressives. Robert Forke, who had resigned as head of the shattered party, called for a union with the Liberals.[16] The U.F.A. candidates, however, voted against any amalgamation and entered the campaign as a separate parliamentary group.[17] The *Grain Growers' Guide* was almost silent on the issues of the 1926 campaign. The Progressive crusade was over, and the *Guide* showed little enthusiasm for the election.

In Alberta alone was the old enthusiasm sustained. The U.F.A. entered the campaign determined to carry the cause of reform to Ottawa once again. The date selected for the election was keenly resented by the Alberta party as an effort by "political tricksters

[15] T. W. Bird added to the confusion by forgetting that his vote was paired. He voted against the government "inadvertently," but the speaker of the house ruled the vote must stand!

[16] Mr. Forke entered the King cabinet as minister of immigration following the election.

[17] *Calgary Daily Herald*, July 24, 1926.

in Ottawa" to place the election in the very midst of harvest "with the premeditated aim of reducing the U.F.A. vote." [18] The U.F.A. representation in Ottawa was increased by the election, however, from nine to eleven.[19] H. W. Wood hailed the election returns as a vindication of the "group government" theory, and the editor of the *U.F.A.* boasted, "In the history of the farmers' political movements the achievements of the Alberta farmers is without precedent." This may seem an extreme statement, but it is substantiated by a survey of all the farmers' movements in the United States and Canada.

The agrarian revolt was over, and the National Progressive movement had passed from the political arena. The vast army of militant farmers was demobilized; only pockets of political resistance remained. Canadian politics seemed fated to return to the traditional two-party system. The future, however, did not reveal so simple a pattern, and the return of economic distress on the prairies brought a new harvest of agrarian radicalism. The seeds of this new and more radical harvest had been planted in the springtime of the Progressive movement. In many respects the heritage bequeathed to the West by the Progressive movement was as important as the revolt itself.

As John W. Dafoe, the editor of the *Manitoba Free Press* and "mentor of Progressivism," surveyed the wreckage of the once powerful Progressive movement he was inspired to write: "The Progressive movement represented a Western outlook, which has not vanished by any means. If it does not present itself through the media of the Progressives it will appear in some other form. That point should not be lost sight of." [20] When agricultural prices collapsed in 1929, and the Great Depression cast its lengthening shadow over the prairies, this shrewd analysis proved to be an accurate prediction of things to come in the Canadian West.

The agrarian revolt on the prairies was generally interpreted as an expression of rural ignorance and rural isolation—a typical frontier phenomenon. Events soon proved it was far more than that. Most eastern observers confidently expected that as the West "set-

[18] *U.F.A.*, 5:6 (September 7, 1926).

[19] E. J. Garland was not one of those who returned. He bolted the U.F.A. to run as a Conservative candidate for Bow River.

[20] *Manitoba Free Press*, October 30, 1925.

tled down" and as it matured, the grain growers would give up their radicalism. Superficial analysis seemed to indicate that this would actually be the case, for as time passed there was no question but that the West did develop a social maturity not present in the initial years of the agrarian revolt. The bitter condemnations of the western farmer, so common during these years, were largely based on the belief that western communities, though built upon a different economy, should become the image of the East.

The collapse of progressivism was accompanied by confident prophecies that the West had reached its maturity — it had come of age. Politicians who should have known better were betrayed by this analysis of prairie life into believing that Canadian politics would no longer be troubled with third parties. Most Canadians predicted that the West had grown out of its callow youth, that it would abandon its blind hate of wealth and privilege and give up its wild criticism of St. James Street.

Rather than this, however, prairie radicalism became more violent than ever. Economic doctrines that seemed at first glance to be strange and new swept the West. This led to a new interpretation of the prairies. Perhaps the "nature" of the people was responsible for their refusal to accept things as they were, or perchance it was because "so many Americans and foreigners" lived in the West.[21] Yet analysis showed that the new leaders were all Canadian or British born. Most observers were willing to blame it on the depression, and since there was no question about this, it sufficed.

The commercialization of agriculture had made the farmer as keenly conscious of money and credits as the businessman or financier. His welfare depended upon a solution to these problems that was favorable to the agricultural economy. Some, of course, advised him to alter his economy, but that required the shaping of factors of climate, soils, and rainfall over which he had slight control. Agrarian discontent, which seemed to die out with the passing of the Progressive party, was only dormant during the

[21] The percentage of foreign born was highest in Alberta and second in Saskatchewan. The census of 1921 gave Alberta 29.56 percent foreign born and Saskatchewan 26.31 percent. The percentage of American born was 16.97 percent in Alberta and 11.57 percent in Saskatchewan. (Canada Bureau of Statistics, *Origin, Birthplace, Nationality and Language of the Canadian People* [Ottawa, 1929], 91, 94.)

favorable years from 1925 to 1929. When it reappeared under the stimulus of depression and agricultural distress, it came garbed in a new dress and waving new banners, but most of the slogans were the same. They were similar because the fundamental demands were similar: favorable marketing prices, social control of currency and credit, and the expansion of cooperative enterprises. Though the movement now entered a formal alliance with labor, this should not conceal the continuity of protest that lived through the sweep of third parties.

Progressive influence in the prairie governments survived the collapse of the national party. The U.F.A. government continued its rule in Alberta until it fell in 1935. In that year, burdened by economic distress it was powerless to check, a victim of unhappy scandals in its official family, palsied by its years in power, and the object of distrust by urban and small-town voters who felt politically dispossessed under the farmers' government, the U.F.A. government fell of its own weight. In Manitoba the U.F.M. government also vanished, but saved much of itself by uniting with the Liberal party. The serious financial burdens of the province during the depression and the pressing problems of unemployment and agricultural distress furnished the Bracken government with the opportunity to appeal for a coalition government. In the election of 1932 the Liberal-Progressive alliance swept the province and progressivism entered the larger life of the Liberal party where its contributions were of significant proportion.

Progressives in Saskatchewan also influenced the older parties in the years following the collapse of the movement. Many Progressives drifted back into the Liberal party as they had in Manitoba. A more important and unusual development was the alliance which many Progressives made with the Conservatives to form an opposition party in Saskatchewan. Dissatisfaction with the Liberal government was so great that in 1929 it was forced from office. The refusal of the United Farmers of Canada to enter politics in that year left no alternative but to elect a predominantly Conservative government. This explains what otherwise would appear to be an anomaly. The "coalition government" under Dr. J. T. M. Anderson ruled the province for four years, when it, too, fell victim to the demands for a change which voters hoped would bring them out of economic depression. Thus, in each of the prairie provinces, the Progressives continued to influence the course of political developments though their own movement had lost its identity.

A more important aspect of the Progressive heritage on the prairies was the underlying change wrought in prairie politics by the movement. The "western outlook" had crystallized into the Progressive movement, and the doctrines of Progressivism remained at political bedrock on the prairies. The conditions that had prompted the Progressive movement remained, and the precedent furnished by the movement was an influential guidepost in the West. The fundamental economic problems that Progressives had wrestled with remained unsolved, as the collapse of world trade in 1929 tragically revealed.

The revolt against "partyism," so much a part of the agrarian crusade, had destroyed the traditional patterns of political action in the West. Prairie voters never returned to their old loyalty to the established parties. The Progressive movement had freed them from the hold of the ancient party lines. The movement had established the precedent of looking to a third party for remedial legislation. Equally important, western farmers were convinced that the older parties were often corrupt and unduly responsive to eastern and industrial influences. A strong strain of political independence was fastened upon the West, and no amount of wishful thinking could restore the western farmers to traditional patterns.

The agrarian revolt had emphasized economic reform. Cooperatives were an integral part of the movement, and the Canadian West possessed a deep and abiding faith in economic doctrines which, in more conservative regions, were regarded with fear and suspicion. The influence of the socialistic beliefs of many of the early grain growers' leaders had created a powerful socialist sentiment which could be fanned into activity by the harsh winds of economic distress. The constant emphasis upon currency and credit manipulation made a deep impression in the West. The agitation by the monetary experts in the movement, especially in the U.F.A., prepared the ground for later doctrines. The demand for easy credit and soft money was a heritage that made the way smooth for later prophets of the new dispensation.

The agrarian revolt taught the prairie farmer to act along class or group lines. This indoctrination was especially strong in the Nonpartisan League and the U.F.A., but it was shared by all the societies in the West. Class differences as well as sectional differences were emphasized throughout the crusade, and a deep distrust of the banker, the industrialist, and the merchant remained after the Progressive movement disappeared. Here was fertile soil for further cultivation by later radicalisms.

Even the word "Progressive" possessed a magic as the result of the farmers' crusade. The psychological value of the term was well recognized by politicians in the West who continued to use it as an appeal to the electorate. The word "Progressive" was capable of arousing old loyalties and attracting new support, and was a mantle of liberalism that could be thrown over nearly any entry in the race. So powerful a magnet did the word become that even the Conservatives were impelled to adopt the anomalous, but politically practical, term "Progressive-Conservative." There were many practical recognitions of the underlying influence and political heritage of the Progressive movement.

The growth of the Cooperative Commonwealth Federation and Social Credit parties revealed the extent of Progressive influence on prairie politics. The Social Credit party in Alberta owed a debt to the U.F.A. which it could hardly be expected to acknowledge. The U.F.A. premier, Mr. Reid, secured the services of Major Douglas as an adviser to the provincial government in 1935. Douglas later pointed out the irony of the U.F.A. defeat at the hands of "Social Credit ideas which they themselves had been instrumental in introducing into the Province." [22]

The U.F.A. prepared Alberta for the new party though it did not plan to do so. The earlier agitation for credit reform by J. W. Leedy, George Bevington, Henry Spencer, William Irvine, and others in the U.F.A. made the transition from cooperative credit to social credit an easy one for Alberta farmers. The basic doctrine that credit controls should rest with the state was well accepted among U.F.A. members long before the Social Credit League began its agitation. The Social Crediters were reaping where others had sown.

The refusal of the U.F.A. to espouse the Social Credit cause

[22] C. H. Douglas, *The Alberta Experiment*, 21.

closed the door on all hopes of absorbing the new movement. It opened the way, however, for the evangelistic pleas of William Aberhart. This schoolmaster-preacher forsook the halls of his Prophetic Bible Institute to preach an economic gospel which, though it bore a new label, had a familiar ring to Albertans. Aberhart's denunciations of St. James Street, his unmeasured attacks on the older parties, and his constant emphasis upon the inadequate and unjust return to the West fell upon receptive ears. Alberta farmers had earlier learned these lessons in the Nonpartisan League and the U.F.A. Added to these old and familiar themes, moreover, was the new and irresistible promise of a monthly dividend of twenty-five dollars for every adult in the province. In the fever pitch of the Social Credit crusade the aging U.F.A. government was doomed. Its defeat was so complete that it won not a single seat in the House. Like the "Deacon's Masterpiece,"

> *It went to pieces all at once,*
> *All at once, and nothing first,*
> *Just as bubbles do when they burst.*

It should be pointed out, however, that if Social Credit had not appeared at that moment to challenge the U.F.A. some other party would have risen. The probabilities are that this other party would have preached much the same gospel, and would have leveled the same indictment against the economic system. The past made that inevitable.

The growth of the Cooperative Commonwealth Federation in Saskatchewan offered further evidence of the deep-rooted heritage of Progressivism. When the United Farmers of Canada began their agitation in 1929 for political action, they were following a familiar course. The resistance to political action in the society gradually died out under the impact of the distress of the early thirties, and a new party was born. The new farmer-labor party was the result of the union of the U.F.C. and organized labor in the province.

The C.C.F., like the Social Credit party, made appeals to passions that were deep rooted in the prairies. Western voters understood what William Irvine meant when he urged them to "reject the old parties in their impotence" for this was an appeal hallowed by years of repetition. Again, when J. S. Woodsworth urged adherence to the C.C.F. program on the grounds that neither of the old parties could be expected to fight the evils in society, he was expressing a conviction that the agrarian revolt had planted years before.

The C.C.F. seemed to be the party the farmers' organizations longed for, now that their own Progressive movement had disappeared. The U.F.A., the U.F.O., and the U.F.M. hastened to endorse the new party. Even the faces were familiar in the new revolt. J. S. Woodsworth, William Irvine, Norman Priestley, Robert Gardiner, Alfred Speakman, W. C. Good—these and many others had fought the good fight in the earlier crusade and were now continuing it under a new banner. That the C.C.F. policy of confiscation alienated many of these leaders and turned the organized farmers against the movement should not conceal the essential continuity of agrarian radicalism in the West. The 1933 "Thirteen Point Manifesto" of the C.C.F. contained little that had not appeared in some phase of the earlier agrarian revolt. Prairie socialism, C.C.F. style, was the logical outgrowth of three decades of agrarian agitation. Now formally allied with labor, it challenged the same powers and recited the same grievances that had been so much a part of the Progressive movement.

The Progressive movement went the way of third parties. Like so many of its predecessors, it left its influence deeply stamped upon the surviving parties. The indoctrination carried on by the agrarian crusade in the early years of the century continued to bear fruit long after the original movement had disappeared. By the time the Great Depression hit the prairie provinces, the Progressive revolt was a thing of the past, but much of its program and most of its doctrines found new life under strange banners. To quote the much later words of the man who first led and then deserted it, this was the movement "that more than any other single thing left a definite and lasting impress upon the political life of the country."[23]

[23] T. A. Crerar, *Winnipeg Free Press*, December 9, 1946.

Bibliography

Unpublished Works

BAHMER, ROBERT H. "The Economic and Political Background of the Nonpartisan League." Ph.D. thesis, University of Minnesota, 1941.

Memorandum to author from John S. Wood, St. Louis, Missouri concerning his brother, the late Henry Wise Wood. Contains reminiscences of John S. Wood and letters from intimate friends of Henry Wise Wood.

PROSKIE, JOHN. "Financial Progress of Settlers with Special Reference to the Vulcan-Lomond Area." M.A. thesis, University of Alberta, 1935.

SALOUTOS, THEODORE. "Farmers' Movements since 1902." Ph.D. thesis, University of Wisconsin, 1940.

WILCOX, BENTON H. "A Reconsideration of the Character and Economic Basis of Northwestern Radicalism." Ph.D. thesis, University of Wisconsin, 1933.

Wood Papers. Clippings and unpublished manuscripts of Henry Wise Wood, in possession of Mr. Leonard D. Nesbitt, Alberta Wheat Pool, Calgary, Alberta.

Government Documents

DOMINION OF CANADA

Bureau of Statistics. *Origin, Birthplace, Nationality and Language of the Canadian People.* Ottawa, 1929.

House of Commons *Debates,* Session 1907–8 to Session 1927.

House of Commons *Journals,* Session 1923, Appendix 2. "Proceedings of the Select Standing Committee of Banking and Commerce." Ottawa, 1923.

House of Commons *Journals,* Session 1923, Appendix 3. "Proceedings of the Select Special Committee on Agricultural Conditions." Ottawa, 1923.

Report of the Royal Commission on Dominion-Provincial Relations, 1867–1939. Ottawa, 1940.

Sessional Papers, 1921, No. 13. "Report of the Chief Electoral Officer on the Fourteenth General Election for the House of Commons." Ottawa, 1921.

Sessional Papers. Reports of the Department of the Interior, 1897–1900, 1907–10, 1912–14, 1916. Ottawa.

Sessional Papers. Reports of the Minister of Agriculture, 1914–19. Ottawa.

UNITED STATES

OKLAHOMA. *Proceedings of the Constitutional Convention of the Proposed State of Oklahoma.* Guthrie, 1907.

UNITED STATES:

Department of Agriculture, *Annual Reports,* 1910–13, 1920–23. Washington.

Department of Agriculture, *Bulletin* 547. O. B. Jesness and W. H. Kerr, "Cooperative Purchasing and Marketing Organization among Farmers in the United States." Washington, 1917.

Department of Agriculture, *Miscellaneous Publication* No. 15. Alfred Charles True, *A History of Agricultural Extension Work in the United States.* Washington, 1928.

Department of Agriculture, Statistics Division, *Miscellaneous Series, Bulletin* 15. H.

T. Newcomb, "Changes in the Rates of Charge for Railway and Other Transportation Services," revised by Edward G. Ward. Washington, 1901.

House Report 408, 67th Congress, 1st Session. "Report of the Joint Commission of Agricultural Inquiry," Vol. 3, pt. I; "The Agricultural Crisis and Its Causes," Vol. 3, pt. II, "Credit." Washington, 1921.

Senate Document 95, 70th Congress, 1st Session. "Cooperative Marketing." Washington, 1928.

Newspapers and Periodicals

Alberta Non-Partisan (Calgary, Alberta), 1917–19. Semimonthly organ of the Nonpartisan League. Files in Legislative Library, Edmonton.

Calgary Daily Herald (Calgary, Alberta), 1914–27. Files in Calgary Public Library and Legislative Library, Edmonton.

Canadian Annual Review of Public Affairs, 1902–36. An indispensable compilation of public events in Canada. Particularly valuable were the sections on Manitoba, Saskatchewan, and Alberta. University of Minnesota Library.

Canadian Forum (Toronto, Ontario), 1920–27. University of Minnesota Library.

Chicago Daily Tribune, May 12, 1920; May 13, 1920.

Edmonton Bulletin (Edmonton, Alberta), 1907–26. Legislative Library, Edmonton.

Edmonton Journal (Edmonton, Alberta), 1921–26. Legislative Library, Edmonton.

Farmers' Advocate and Home Journal (London, Ontario, and Winnipeg, Manitoba), 1896–1927. University of Manitoba Library.

Grain Growers' Guide (Winnipeg, Manitoba), 1908–27. Weekly organ of the three prairie grain growers' organizations during the period of this study. Complete file in offices of the *Country Guide*, Winnipeg, Manitoba.

Industrial Canada (Toronto, Ontario), 1910–27. Monthly organ of the Canadian Manufacturers' Association. Incomplete files in Legislative Library, Winnipeg, University of Manitoba Library, and University of Minnesota Library.

Lumber World Review (Chicago, Illinois), February 25, 1919. Forest Products History Foundation, Minnesota Historical Society, St. Paul.

Manitoba Free Press (Winnipeg, Manitoba), 1910–27. Minnesota Historical Society Library, St. Paul.

Nonpartisan Leader (Fargo, North Dakota), 1915–19. Official organ of the Nonpartisan League of North Dakota. University of Minnesota Library.

St. Paul Pioneer Press (St. Paul, Minnesota), August 21, 1913.

Nutcracker (Calgary, Alberta), November 1916 to September 1917. Semimonthly independent journal which was a proponent of the Nonpartisan League. Legislative Library, Edmonton.

Sun (Swift Current, Saskatchewan), 1915–21. File in the *Sun* office, Swift Current.

U.F.A. (Calgary, Alberta), 1922–34. Semimonthly organ of the United Farmers of Alberta. File in the *U.F.A.* offices, Calgary.

United Farmer (Calgary, Alberta), 1934–35. Semimonthly continuation of the *U.F.A.* File in *U.F.A.* offices, Calgary.

Western Independent (Calgary, Alberta), October 1919–June 1920. Weekly organ of the U.F.A. Political Association. In June 1920 the U.F.A. executive dropped its financial support. Legislative Library, Edmonton.

Winnipeg Tribune (Winnipeg, Manitoba), 1921–33. Legislative Library, Winnipeg.

Contemporary Articles, Books, and Pamphlets

ANONYMOUS. "The Agrarian Movement in Canada." *Quarterly Review*, 235:84–107 (January 1921).

ANONYMOUS. "Ku Klux Klan in Saskatchewan." *Queens Quarterly*, 35:592–602 (Autumn 1928).

BORDEN, HENRY (editor). *Robert Laird Borden: His Memoirs.* London, 1938.

CANADIAN RECONSTRUCTION ASSOCIATION. "Agriculture and Industry," Toronto, 1920; "The Nonpartisan League in North Dakota, a study of a class war and its disastrous consequences," Toronto, 1919; "The Outlook for Canada," Toronto, 1920; "Ways to National Prosperity," Toronto, 1920.

CARP, C. L. "Fate of the Grain Growers." *Alberta Non-Partisan*, September 27, 1918.
CHIPMAN, GEORGE F. *The Siege of Ottawa.* 68 pages. Winnipeg, 1910.
CLAY, H. C. "Reminiscences." *Grain Growers' Guide*, 9:21 (March 1, 1916).
CRERAR, THOMAS A. "My Confession of Faith." *Grain Growers' Guide*, 14:7 (February 23, 1921).
DAFOE, JOHN W. "Western Canada: Its Resources and Possibilities." *Review of Reviews*, 35:697–710 (June 1907).
DIXON, F. J. "Independent Political Action Needs." *Grain Growers' Guide*, 8:8 (September 9, 1915).
FARMER, SEYMOUR J. "Direct Legislation in Manitoba." *Grain Growers' Guide*, Vol. 4, December 6, 1911.
"The Farmers' Platform." 54 pages. Winnipeg, 1917.
FRASER, JOHN FOSTER. *Canada As It Is.* London, 1911.
GODWIN, SIDNEY. "The Farmers' Political League." *Grain Growers' Guide*, 9:19 (October 4, 1916).
GOOD, W. C. "The Farmer's Movement in Canada." *Dalhousie Review*, 3:476–84 (January 1923).
———. *Production and Taxation in Canada from the Farmers' Standpoint.* Toronto, 1919.
GUSTUS, J. E. "Alfalfa — A Prosperous Yankee Settler." *Grain Growers' Guide*, March 27, 1912.
HASLAM, J. H. "Why Settlers Leave." *Grain Growers' Guide*, 9:8, 19, 20 (August 30, 1916).
HOPKINS, J. CASTELL. "Canada's Conservative Policies." *North American Review*, 194:818–25 (December 1911).
HOWE, FREDERICK C. "Oregon: The Most Complete Democracy in the World." *Grain Growers' Guide*, 3:7 (May 3, 1911).
HUFFMAN, BERT. "A Harvest of Golden Principles, or a Harvest of Chaff." *U.F.A.*, 4:4 (October 24, 1925).
HURD, ARCHIBALD S. "Foreign Invasion of Canada." *Fortnightly Review* (London), 78:1055–65 (December 1902).
IRVINE, WILLIAM. *The Farmer in Politics.* Toronto, 1921.
LAMBERT, NORMAN P. *The Farm Movement in Canada.* Toronto, 1922.
"Land Hungry Go to Canada." *Nonpartisan Leader* (Fargo, North Dakota), 5:11 (October 25, 1917).
LAUT, AGNES C. *Canada at the Crossroads.* Toronto, 1921.
———. "The Last Trek to the Last Frontier." *Century Magazine*, 78:99–112 (May 1909).
LEACOCK, STEPHEN. "Our Organization for War." In John O. Miller (editor), *The New Era in Canada.* London, 1917.
"A League in Canada." *Nonpartisan Leader* (Fargo, North Dakota), 4:8 (February 8, 1917).
MANITOBA GRAIN GROWERS' ASSOCIATION. Pamphlets, bulletins, brochures, mimeographed material, newspaper clippings in boxes in the *Country Guide* library, Winnipeg, Manitoba.
———. *Yearbooks*, 1913–19.
McKENZIE, RODERICK. "How Farmers Captured a State." *Grain Growers' Guide*, 9:18 (September 6, 1916).
MILLER, JOHN O. (editor). *The New Era in Canada: Essays Dealing with the Upbuilding of the Canadian Commonwealth.* London, 1917.
MOORHOUSE, HOPKINS, *pseud.* (Herbert Joseph). *Deep Furrows.* Toronto, 1918.
MORTON, JAMES. "Farmers in Politics." *Grain Growers' Guide*, 7:13 (August 5, 1914).
MYERS, GUSTAVUS. *History of Canadian Wealth.* Chicago, 1914.
NELSON, JOHN. *The Canadian Provinces: Their Problems and Policies.* Toronto, 1924.
"The New National Policy as Adopted by the Organized Farmers of Canada in the Farmers' Platform." 20 pages. Winnipeg, 1920.
"The Nonpartisan League." *Grain Growers' Guide*, 10:8, 24 (September 5, 1917).
NORTH, NICHOLAS. "That Man from Carstairs." *Maclean's*, 35:27 (February 1, 1922).
O'DONNELL, JOHN H. *Manitoba As I Saw It from 1869 to Date.* Toronto, 1909.

PACKARD, FRANK. "Tough on Speculators, How the Northwestern Provinces of Canada Have Lightened the Burdens of the Producing Farmer by Taxing Idle Land." *Nonpartisan Leader* (Fargo, North Dakota), 2:3 (March 9, 1916).

PATRONS OF HUSBANDRY. *Journal of the Proceedings of the Dominion Grange.* 1909.

PETERSON, CHARLES W. *Wake Up, Canada: Reflections of Vital National Issues.* Toronto, 1919.

"Political and Constitutional Crisis." *Round Table,* 16:824–36 (September 1926).

PORRITT, EDWARD. *The Revolt in Canada against the New Feudalism.* London, 1911.

SASKATCHEWAN GRAIN GROWERS' ASSOCIATION. Bulletins, pamphlets, newspaper clippings, mimeographed material in boxes in the *Country Guide* library, Winnipeg.

SHARP, WALTER R. "Canadian Elections of 1925." *American Political Science Review,* 20:107–17 (February 1926).

———. "Canadian Election of 1926." *American Political Science Review,* 21:101–13 (February 1927).

SMITH, GOLDWIN. *Reminiscences.* New York, 1910.

SMITH, W. L. "When the Patrons Swept Ontario." *Grain Growers' Guide,* 19:6 (December 1, 1926).

STEVENSON, J. A. "The Agrarian Movement in Canada." *Edinburgh Review,* 231:96–113 (July 1920).

———. "The Battle for Democracy in Canada." *Grain Growers' Guide,* 3:7 (November 2, 1910).

———. "The Future of Western Democracy." *Grain Growers' Guide,* 9:8, 24 (July 12, 1916).

———. "The Political and Economic Situation in Canada." *Edinburgh Review,* 241:209–29 (April 1925).

———. "The Political Situation in Canada." *Contemporary Review* (London), 129:444–61 (April 1926).

———. "Strains on Canadian Confederation." *New Republic,* 41:166–68 (January 7, 1925).

THOMPSON, BRAM. *Canada's Suzerainty over the West: An Indictment of the Dominion and Parliament of Canada for the National Crime of Usurping the Public Land of Manitoba, Saskatchewan and Alberta.* Toronto, 1919.

THWAITE, LEO. *Alberta, An Account of Its Wealth and Progress.* London, 1912.

UNITED FARMERS OF ALBERTA. *Annual Report and Yearbook,* 1917–27.

———. "U.F.A. Provincial Platform, Declaration of Principles." Undated (1921).

UNITED FARMERS OF MANITOBA. Bulletins, newspaper clippings, political campaign pamphlets, and mimeographed material in boxes in the *Country Guide* library, Winnipeg, Manitoba.

———. *Handbook of Practical Work.* Winnipeg, undated.

———. "Some Pertinent Facts about the U.F.M." Winnipeg, 1925.

WALKER, B. E. "The East and West." In John O. Miller (editor), *The New Era in Canada.* London, 1917.

WELLIVER, JUDSON C. "The Agricultural Crisis and the 'Bloc.'" *Review of Reviews,* 65:158–65 (February 1922).

WEST, EDWARD. *Homesteading: Two Prairie Seasons.* London, 1920.

"Why Our Farmers Seek Canada." *Literary Digest,* 45:1217–19 (December 28, 1912).

WILLISON, J. S. "Immigration and Settlement." In John O. Miller (editor), *The New Era in Canada.* London, 1917.

WILSON, JOHN. "Experiences of a Homesteader." *Grain Growers' Guide,* 4:44–45 (December 6, 1911).

WOOD, HENRY WISE. "The Canadian Council of Agriculture and Political Action." *U.F.A.,* 2:9 (May 15, 1923).

———. "The Efficient Citizenship Group." *Grain Growers' Guide,* 15:7 (March 22, 1922).

———. "In Defense of Group Politics." *Canadian Forum,* 3:72 (December 1922).

———. "The Group versus the Party." *Grain Growers' Guide,* 15:17 (September 6, 1922).

———. "The Organized Farmers and Politics." *Grain Growers' Guide,* 10:10 (September 19, 1917).

———. "Pendulum of the Clock without Hands Makes Another Swing." *U.F.A.*, 4:1, 3 (November 2, 1925).
———. "The Significance of Democratic Group Organization." *U.F.A.*, 1:5 (April 15, 1922).

Secondary Works

ANGUS, HENRY FORBES (editor). *Canada and Her Great Neighbor.* New Haven, 1938.
BAHMER, ROBERT H. "The Society of Equity." *Agricultural History,* 14:33–63 (January 1940).
BAKKEN, HENRY HARRISON, and MARVIN A. SCHAARS. *Economics of Cooperative Marketing.* New York, 1937.
BOYLE, JAMES E. "The Agrarian Movement in the Northwest." *American Economic Review,* 8:506–21 (September 1918).
BRITNELL, G. E. *The Wheat Economy.* Toronto, 1939.
BUCHANAN, DONALD W. "The Mormons in Canada." *Canadian Geographical Journal,* 2:255–70 (April 1931).
BULLER, A. H. REGINALD. *Essays on Wheat.* Toronto, 1919.
CLARK, S. D. "The Canadian Manufacturers' Association: A Political Pressure Group." *Canadian Journal of Economics and Political Science,* 4:505–23 (November 1938).
COATS, ROBERT HAMILTON, and MURDOCH C. MACLEAN. *The American-Born in Canada: A Statistical Interpretation.* New Haven, 1943.
CURRIE, A. W. "Freight Rates on Grain in Western Canada." *Canadian Historical Review,* 21:40–55 (March 1940).
CRERAR, THOMAS A. "Wheat: The Early History." *Winnipeg Free Press,* December 9, 1946.
———. "Wheat: Looking to Stability." *Winnipeg Free Press,* December 11, 1946.
DAFOE, JOHN W. *Clifford Sifton in Relation to His Times.* Toronto, 1931.
DESTLER, CHESTER MCARTHUR. "Western Radicalism, 1865–1901: Concepts and Origins." *Mississippi Valley Historical Review,* 31:335–68 (December 1944).
DAWSON, ROBERT MACGREGOR (editor). *Constitutional Issues in Canada, 1900–1931.* London, 1933.
DOUGLAS, CLIFFORD HUGH. *The Alberta Experiment: An Interim Study.* London, 1937.
EASTERBROOK, W. T. *Farm Credit in Canada.* Toronto, 1938.
ELLIS, L. ETHAN. *Reciprocity of 1911.* New Haven, 1939.
ENGLAND, ROBERT. *The Colonization of Western Canada: A Study of Contemporary Land Settlement (1896–1934).* London, 1936.
FALCONER, SIR ROBERT A. *The United States as a Neighbor, from a Canadian Point of View.* Cambridge, 1925.
FAULKNER, HAROLD UNDERWOOD. *The Quest for Social Justice, 1898–1914.* New York, 1931.
FORD, ARTHUR R. "Some Notes on the Formation of the Union Government in 1917." *Canadian Historical Review,* 19:357–64 (December 1938).
FOSSUM, PAUL R. "Early Milling in the Canon River Valley." *Minnesota History,* 11:271–82 (September 1930).
FOWKE, VERNON C. *Canadian Agricultural Policy, The Historical Pattern.* Toronto, 1946.
———. "Economic Effects of War on Prairie Economy." *Canadian Journal of Economics and Political Science,* 11:373–87 (August 1945).
GLAZEBROOK, GEORGE PARKIN DE TWENEBROKER. *A History of Transportation in Canada.* New Haven, 1938.
GRANT, H. C. "The Cooperative Movement: Developments in the Concept of Cooperation in Western Canada." *Canadian Journal of Economics and Political Science,* 3:412–20 (August 1937).
HANSEN, MARCUS LEE, and JOHN BARTLETT BREBNER. *The Mingling of the Canadian and American Peoples.* New Haven, 1940.
HICKS, JOHN D. *The Populist Revolt.* Minneapolis, 1931.
———. "The Western Middle West, 1900–1914." *Agricultural History,* 20:65–77 (April 1946).
INNIS, MARY Q. *An Economic History of Canada.* Toronto, 1935.

KEENLEYSIDE, HUGH LLEWELLYN. *Canada and the United States.* New York, 1929.

KILE, ORVILLE M. *The Farm Bureau Movement.* New York, 1921.

KUHLMAN, CHARLES BYRON. *The Development of the Flour Milling Industry in the United States.* New York, 1929.

MACGIBBON, DUNCAN A. *The Canadian Grain Trade.* Toronto, 1932.

MACKINTOSH, WILLIAM ARCHIBALD. *Agricultural Cooperation in Western Canada.* Toronto, 1924.

———. *Economic Problems of the Prairie Provinces.* Toronto, 1935.

———. *Prairie Settlement, The Geographical Setting.* Toronto, 1934.

———. "Some Aspects of a Pioneer Economy." *Canadian Journal of Economics and Political Science,* 2:457–63 (November 1936).

MCGOUN, A. D. "Alberta, Economic and Political." *Canadian Journal of Economics and Political Science,* 2:512–24 (November 1936).

MCWILLIAMS, MARGARET. *Manitoba Milestones.* Toronto, 1928.

MORMAN, JAMES B. *Farm Credits in the United States and Canada.* New York, 1924.

MORTON, W. L. "Direct Legislation and the Origins of the Progressive Movement." *Canadian Historical Review,* 25:279–88 (September 1944).

———. "The Western Progressive Movement, 1919–1921." *Proceedings* of the Canadian Historical Association, 1946, 41–55.

———. "Western Progressive Movement and Cabinet Domination." *Canadian Journal of Economics and Political Science,* 12:136–47 (May 1946).

MURCHIE, ROBERT WELCH. *Agricultural Progress on the Prairie Frontier.* Toronto, 1936.

PARRINGTON, VERNON L. *The Beginnings of Critical Realism.* New York, 1930.

PATTON, HARALD S. *Grain Growers' Cooperation in Western Canada.* Cambridge, 1928.

———. "Observations on Canadian Wheat Policy Since the World War." *Canadian Journal of Economics and Political Science,* 3:218–32 (May 1937).

POWELL, G. HAROLD. *Cooperation in Agriculture.* New York, 1918.

PRINGLE, HENRY F. *The Life and Times of William Howard Taft.* New York, 1939.

REID, ESCOTT. "Saskatchewan Liberal Party before 1929." *Canadian Journal of Economics and Political Science,* 2:27–40 (February 1936).

SALOUTOS, THEODORE. "Expansion and Decline of the Nonpartisan League in the Western Middle West, 1917–1921." *Agricultural History,* 20:235–52 (October 1946).

———. "The Montana Society of Equity." *Pacific Historical Review,* 14:393–408 (December 1945).

———. "The Rise of the Equity Cooperative Exchange." *Mississippi Valley Historical Review,* 32:31–62 (June 1945).

———. "The Rise of the Nonpartisan League in North Dakota, 1915–1917." *Agricultural History,* 20:43–61 (January 1946).

SHANNON, FRED A. *The Farmer's Last Frontier: Agriculture, 1860–1897.* New York, 1945.

———. "A Post Mortem on the Labor-Safety-Valve Theory." *Agricultural History,* 19:31–37 (January 1945).

SHARP, MITCHELL W. "Allied Wheat Buying in Relationship to Canadian Marketing Policy, 1914–18." *Canadian Journal of Economics and Political Science,* 6:372–89 (August 1940).

STEWART, ANDREW. *The Economy of Machine Production in Agriculture.* Montreal, 1931.

TANSILL, CHARLES C. *Canadian-American Relations, 1875–1911.* New Haven, 1943.

UNDERHILL, FRANK H. "The Party System in Canada." *Proceedings* of the Canadian Political Science Association, 1932, 201–12.

WOOD, LOUIS AUBREY. *A History of the Farmers' Movements in Canada.* Toronto, 1924.

Index